Movements of the Hand

The Sherrington Lectureship

In 1948, on the occasion of the ninetieth birthday of Sir Charles Scott Sherrington, OM, GBE, MD, FRS, the Council of the University of Liverpool resolved to institute a Lectureship in recognition of his distinguished contributions to Physiology and Medicine and of his association with the University as George Holt Professor of Physiology from 1895 to 1913. The appointment to the Sherrington Lectureship is made biennially by the Council of the University on the joint recommendation of the Faculties of Medicine and Science.

The Sherrington Lectures

The following Lectures have been delivered and, except as indicated, are published by Liverpool University Press.

I *Sensory Integration*, by E. D. Adrian, OM, MD, FRCP, FRS, Professor of Physiology in the University of Cambridge.

II *The Frontal Lobes and Human Behaviour*, by John F. Fulton, OBE, MD, DSC, LLD, Sterling Professor of the History of Medicine, Yale University.

III *The Invasive Adenomas of the Anterior Pituitary*, by Sir Geoffrey Jefferson, MS(LOND), FRCS, FRS, Emeritus Professor of Neurosurgery, University of Manchester.

IV *Sherrington: Physiologist, Philosopher and Poet*, by the Right Honourable Lord Cohen of Birkenhead, MD, DSC, LLD, FRCP, FACP, FFR, FSA, JP, Professor of Medicine in the University of Liverpool.

V *The Excitable Cortex in Conscious Man*, by Wilder Penfield, OM, CMG, LITTB, MD, DSC, FRCS, HON FRCP, FRS, Director, Montreal Neurological Institute, McGill University, Montreal, Canada.

VI *Visual Pigments in Man*, by W. A. H. Rushton, SCD, FRS, Reader in Physiology in the University of Cambridge.

VII *The Conduction of the Nervous Impulse*, by A. L. Hodgkin, SCD, FRS, Foulerton Research Professor of the Royal Society, University of Cambridge.

VIII *The Cerebral Control of Movement*, by Derek Denny-Brown, OBE, MD, DPHIL, LLD(HON), James Jackson Putnam Professor of Neurology, Harvard University.

IX *The Inhibitory Pathways of the Central Nervous System*, by John C. Eccles, FRS, MB, BS, MA, DPHIL, FRACP, FRSNZ, FAA, Faculty of Health Services, State University of New York at Buffalo.

X *The Release of Neural Transmitter Substances*, by Bernard Katz, MD, DSC, FRS, Professor of Biophysics, University College London.

XI *Mechanisms Regulating the Discharge of Motoneurons*, by Ragnar Granit, PHILMAG, MD, DSC, FORMERRS, Professor Emeritus of Neurophysiology, The Royal Caroline Institute, Stockholm, Sweden.

XII *Synaptic Transmission: A Microphysiological Analysis*, by Stephen W. Kuffler, MD, FORMEMRS, Robert Winthrop Professor of Neurobiology at the University of Harvard. *Not published.*

XIII *The Parietal Lobe and Extrapersonal Space*, by Vernon B. Mountcastle, MD, Professor of Physiology, The Johns Hopkins University School of Medicine, Baltimore. *Not published.*

XIV *Reflections on Muscle*, by Sir Andrew Huxley, SCD, FRS, Royal Society Research Professor, Department of Physiology, University College London.

XV *Release and Uptake Functions in Adrenergic Nerve Granules*, by Ulf S. von Euler, MD, FORMEMRS, Professor Emeritus of Physiology in The Royal Caroline Institute, Faculty of Medicine, Stockholm, Sweden.

XVI *Fifty Years On: Looking back on some developments in Neurohumoral Physiology*, by W. S. Feldberg, CBE, MD, FRCP, FRS, Emeritus Professor, Former Head of the Laboratory of Neuropharmacology, National Institute for Medical Research, London.

XVII *Movements of the Hand*, by C. G. Phillips, FRS, Dr Lee's Professor-emeritus of Anatomy in the University of Oxford, Emeritus Fellow of Trinity and Hertford Colleges.

XVIII *Hypothalamic Control of Pituitary Functions. The Growth Hormone Releasing Factor*, by Roger Guillemin, Laboratories for Neuroendocrinology, The Salk Institute for Biological Studies, La Jolla, California, USA.

The Sherrington Lectures XVII

Movements of the Hand

C. G. Phillips, FRS

Dr Lee's Professor-emeritus of Anatomy
in the University of Oxford
Emeritus Fellow of Trinity and Hertford Colleges

LIVERPOOL UNIVERSITY PRESS

Published by
LIVERPOOL UNIVERSITY PRESS
PO Box 147, Liverpool L69 3BX

ISBN 0 85323 095 1

First published 1985

British Library Cataloguing-in-Publication Data

Phillips, C. G.
Movements of the hand.—(The Sherrington lectures; 17)
1. Human mechanics 2. Hand
I. Title
612′.76 QP301

ISBN 0–85323–095–1

Text set in 10/12 pt Linotron Times
Printed and bound in Great Britain
at The Bath Press, Avon

Dedication

To the people who have so kindly given me their typescripts, offprints and even monographs, over the years.

Preface

In thanking the University of Liverpool for inviting me to give the XVII Sherrington Lecture on 24 November 1982, I cannot forget that the honour belongs at least as much to the many friends with whom I have been privileged to do research as it does to myself.

I took the invitation as an encouragement to try to design a lecture for medical people without much professional interest in neuroscience at its more Rosicrucian levels. I hope that no-one in the audience regretted this approach. I wanted to disengage as far as possible from the details of experiments in which I have shared; to highlight some of the problems that are still with us, in some cases nearly two centuries after their first recognizable formulation; and to show how some of the formulations have been given greater precision as new experiments have made this possible. In particular I wanted to show the continuing fertility of some of Sherrington's seminal discoveries and concepts. These motives, rather than any wish to review some narrower field of neuroscience in greater depth, determined the structure not only of the Lecture itself but also of its lecture-style expansion in this book. What might otherwise have been 'small print' sections have been relegated to the Notes at the end. The narrative is uninterrupted by proper names—almost every sentence would have needed one at least—but the Notes refer to enough literature to encourage readers to find out more about experiments that I hope may have intrigued them, and above all to expose the many places where I have been skating on thin ice, and making things sound so much simpler than they really are.

It is a pleasure to thank Jane Ballinger for her expert and cheerful work at the word-processor, and Brian Archer and Ken Collins for their work on the illustrations. Much of the final re-writing was done during the tenure of a Visiting Fellowship at the John Curtin School of Medical Research, The Australian National University, Canberra, and I thank the Director, Professor Robert Porter FAA, for all the facilities he so generously made available to me. Rosalind Campbell, Secretary and Publisher at Liverpool University Press, Robin Bloxsidge, Assistant Secretary, and Patricia Wood gave me every help at all stages of the publication process.

C.G.P.

Preface

[illegible]

Contents

V. 'The muscular sense'

VI. 'The muscular sense': perception of passive movements of the fingers

VII. The 'muscular sense': perception of active movements

VIII. Active touch

IX. Cortex and hand

Figures

CHAPTER I

Introduction

1. *Looking, touching and taking hold*

Sherrington's Marshall Hall Prize Address, *On the Spinal Animal* (1899),[1] contains an arresting statement: 'That which the delicate yellow spot is to the sensifacient sheet of the retina, may the thumb and index be said to constitute in the great sensifacient field of the limb.'

In its context the statement refers merely to the virtually total abrogation of spinal reflexes in monkeys in the state of acute spinal shock. Stimulation of the thumb and index finger, the 'yellow spot' of the limp limb, even by 'insults of a character severe in the extreme', evokes either no responsive movement or, at most, a faint flexion limited to the thumb.

But the notion of the primate's thumb and index as the macula of its somaesthetic system, analogous to the macula of its visual system, invites discussion in an altogether wider context, one which is implicit as well as explicit in many of Sherrington's writings. This is that our explorations of tactual and visual space require active movements of the 'sensifacient fields' to centre their respective maculae on objects of interest. In our most delicate performances, as in microsurgery, the finest movements of the fingers must be under simultaneous control from the very centres of the visual and tactual maculae. Some of the foundations for such a wider discussion were firmly laid by Sherrington's classical experiments on the cortical control of hand and eye movements in chimpanzee, orang-utan and gorilla, done during his years at Liverpool.

The comparison of the eye with the hand had been made earlier, by Charles Bell in his Bridgewater Treatise on The Hand.[2] 'There is a small spot opposite to the pupil and in the axis of the eye, which is more peculiarly sensitive to visual impressions . . . There is a continual desire of exercising the sensible spot . . . When an impression is made upon the retina, in that unsatisfactory degree, which is the effect of its striking any part but the centre, there is an effort made to direct the axis towards it, or, in other words, to receive the rays from it upon the more sensible centre. It is this sensibility, therefore, conjoined with the muscles of the eyeball, which produce the constant searching movement of the eye ... This faculty of searching for the object is slowly acquired in the child: and in truth, the motions of the eye are made perfect, like those of the hand by slow degrees. In both organs there is a compound operation: the impression on the nerve of sense is accompanied with an effort of the will, to accommodate the muscular action to it ... On coming into a room, we see the whole side of it at once—the mirror, the pictures, the cornice, the chairs; but we are deceived: being unconscious of the motions of the eye, and that each object is rapidly, but successively, presented to it.' These ideas anticipate the experiments of Yarbus a century-and-a-third later.[3]

In the 1930s Gordon Holmes[4] wondered how a student might respond to the question 'How do we *look* at an object and arrange to see it distinctly? The answer might be given: "I move my eyes by an effort of will as I would move my arm, and I keep the object in central vision by appropriate muscular contractions as I would if it were grasped in my hand."' Holmes went on to show, from clinical research, that visual fixation and binocular fusion, in the pursuit of moving as well as in the holding of stationary objects, are automatic not voluntary actions, sustained by reflexes arising in the retinae and backed up by vestibulo-ocular and optokinetic reflexes which compensate for fast and slow displacements of the head. Fixation is temporarily broken by reflexes from the extra-macular areas of the retina, which automatically attract the gaze towards unexpectedly-moving objects in the peripheral field of vision, and by auditory reflexes which attract it towards the sources of unexpected sounds. Fixation is also suspended during voluntary direction of the gaze. When voluntary control is lost as a result of lesions of the cortex or of its projections to the mid-brain the fixation reflex is exaggerated, and can be broken only by screening the eyes.

Voluntary movements of the hand are also associated with reflexes which may automatically bring the glabrous skin of the finger-pads and palms—the somaesthetic 'macula'—into contact with an object which has touched some other part of the forearm. 'When a spot of less discriminative sensibility is touched, instinct moves the member, so that it brings to the object the part where its own sensitivity is delicate, e.g. the finger tips' (Sherrington, 1900). Denny-Brown's Sherrington Lectures (1966) reviewed his own clinical and experimental studies of the 'instinctive tactile grasp reaction', whereby the hand automatically orientates itself to take hold of an object it has touched, e.g. with the back of the hand: monkeys lose this reaction, as well as voluntary control of the hand, after a lesion of the pyramidal tract. The 'grasp reflex', however, the automatic closure of the fingers upon an object which has been brought into contact with the finger-pads and palm, is exaggerated in the mild flexor spasticity of the fingers which follows a pyramidal lesion. It is not too fanciful to think of the instinctive tactile grasp reaction as analogous to the attraction of the visual axis to an object in the peripheral field, and the grasp reflex as analogous to the fixation reflex.

Instructive comparisons can be drawn between the two sets of musculoskeletal apparatus that bring the visual axis and the thumb and index on to their targets, and between the different mechanical constraints under which they operate. The comparisons should lead us to expect corresponding differences, as well as similarities, in the details of the central nervous apparatus that control the two sets. Common to both is the need for stable postural support. The head is held steady by legs, trunk and neck, controlled by reflexes arising in the retinae and in the vestibular apparatus which is, as it were, bolted to the skull. After bilateral lesions of the vestibular nerves the holding of the skull is too unsteady for accurate fixation to be possible, so that the patient cannot recognize familiar faces. The arm similarly depends on postural support by legs, trunk and shoulder-girdle. Sherrington discovered that very small postural displacements are corrected by the stretch reflexes of antigravity muscles, coordinated by vestibular and visual control. It has been known since Goldscheider that the sensitivity of the shoulder joint to angular movement is greater than that of the finger joints. We know now that small passive displacements of the hand, which do not directly stretch the proximal and axial antigravity muscles that are engaged in maintaining posture, automatically evoke anticipatory postural corrections in these muscles.[5]

Helmholtz drew attention to the fact that when the gaze is redirected automatically or voluntarily there is no apparent movement of the visual field, although the retina has been moved and the images of objects have moved across it; but that when the eyeball is moved passively the field *is* seen to move. He concluded that when the retina is moved by the eye muscles, and not by an experimenter's finger, the perception of movement is annulled by a 'sensation of innervation' aroused by the motor outflow from the brain. So we feel that we

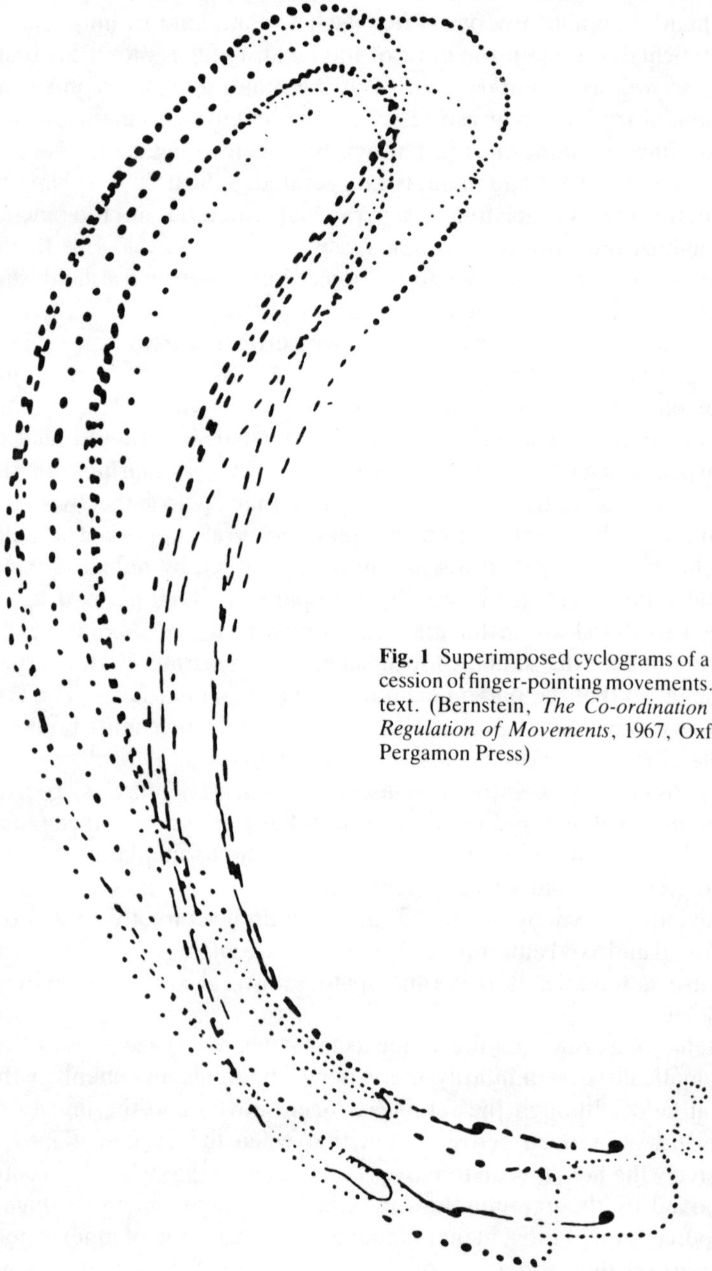

Fig. 1 Superimposed cyclograms of a succession of finger-pointing movements. See text. (Bernstein, *The Co-ordination and Regulation of Movements*, 1967, Oxford: Pergamon Press)

are moving about in a stationary three-dimensional framework of visual space, and that we are directing our gaze, like a probing search-light, from object to object in the stationary framework. Within the same framework we move our arms to probe and grasp the objects that lie within reach of our hands. Our 'body image' is orientated within the same framework, and the images of our arms and hands are orientated to the body image as a whole as well as to the three-dimensional visual, auditory and gravitational frames of reference.

The musculoskeletal apparatus within the orbit provides for movement of each visual axis at a single ball-and-socket joint. The extracorporeal, cyclopean visual axis is an immaterial probe, driven by muscular and elastic forces confined wholly within the orbits. With the head clamped, repeated transfer of the gaze from a fixated point A to a target point B would always have to be brought about by a nearly identical combination and sequencing of the contractions and relaxations of the twelve eye muscles.

In striking contrast, the musculoskeletal apparatus of the arm includes several joints, most of them with more than one degree of freedom; and most of the muscles cross more than one joint. So it would be theoretically possible to move the tip of the index finger from Point A to Point B along *identical* trajectories, using *different* contortions of proximal muscles and joints on every trial. Bernstein [6] defined 'co-ordination of a movement' as 'the process of mastering redundant degrees of freedom of the moving organ, in other words its conversion to a controllable system.' Granit,[7] also, has remarked on the embarrassing wealth of degrees of freedom and the need for 'teleological constraints in order to narrow down the number of possibilities in a sensible manner.' No such extremes of peripheral redundancy embarrass the central control of the visual axis. In the case of the arm, Bernstein concluded from 'cyclogrammetric records of the movement of pointing with a finger at an object, carried out with optimal skill and accuracy,' that 'N successive gestures by the same subject are made through N non-coincident trajectories which only gather, as at a focus, in the vicinity of the same required point which is being indicated.' (Fig. 1). There is also variation in the speedings-up and slowings-down at different parts of the trajectories. But the scatter of the trajectories is relatively slight by comparison with the theoretical possibilities, showing that the central nervous system has indeed successfully mastered the 'redundant degrees of freedom of the moving organ'.

Movements of arm and hand differ further from those of the visual axis in that the heavy proximal segments are constrained by the gravitational field, and that these segments, and also the relatively weightless fingers, may encounter 'unexpected' resistances which interfere with the execution of a programmed movement. And finally, the act of grasping an object includes more than the accurate direction of the movement on to the target: the pressure of the grasp must neither crush an egg nor allow it to slip from the fingers.

Oculomotor physiology has made great strides in the last fifteen years. The neuronal machinery that steers the monkey's gaze as it explores the visual world by coordinated movements of head and eyes, and responds automatically to visual and other stimuli, is now being elucidated in combined behavioural and electrophysiological experiments. The machinery is being modelled in terms of an increasingly sophisticated control systems analysis. The models are grounded on the wealth of data made available by an improving microelectrode technology, and generate predictions that can be tested in fresh experiments. Work on movements of the hand has lagged somewhat behind, but is now showing signs of catching up and attracting more people into this field of research. Studies of the neuronal hardware, and simultaneous recordings of neuronal activities and movements of forearm, hand and fingers, are making good progress, thanks to the introduction of powerful new techniques, some of them applicable not only to monkeys but also to healthy people and to patients with neurological illnesses. The greater degrees of freedom of the movements, as compared with those of the visual axis, will offer a difficult but not ultimately insuperable challenge to the modelling and predictive skills of the control systems analyst.

2. *Movements of the hand*

The subject is vast, and the scope of the Lecture had to be limited to the elemental motor functions of tactile exploration, prehension and manipulation in animals that use their fingers, instead of their muzzles, for these purposes. In their musculoskeletal structure the hands of primates have remained primitive. To appreciate this we have only to compare their pentadactyl hands with the homologous structures in cat and horse. The increasing versatility of hand function as we ascend the primate scale has been conferred by a progressive enlargement of cerebral cortex and cerebellum and a corresponding enrichment of their intrinsic and extrinsic synaptic connectivity.

In Old World Monkeys, Apes and man, Napier[8] has differentiated power and precision grips: power, in use of the whole hand for supporting weight or wielding heavy objects; precision, in picking up small objects with thumb and index finger, whose sensitive tips can be brought into pulp-to-pulp contact by the elemental movements of flexion and of opposition. Using combinations and sequences of the elemental movements of flexion, extension, adduction, abduction and opposition of thumb and index, any baboon can extract the sting from a scorpion and any chimpanzee a thorn from its skin.[9] From the same elemental units of movement, man's highest manipulative skills—drawing, sculpture, writing, watchmaking, microelectronics, microsurgery—are built up. Maximum differentiation of the movements of individual fingers is seen in musicians and touch-typists. When, as beginners, these people were learning to finger the keys, their fingers were at the centre of the field of consciousness; but hard practice eventually displaced the fingers to the fringe of the field, leaving the overriding goals of performance—musical or linguistic—in possession of the centre.

But the Lecture had to content itself with the neurophysiological and neuropsychological elements of manual motor performance, not with the summits of human skill. It had two leading themes.

First, experiments on how we know the positions and movements of our fingers and on how we control them when picking up and exploring objects by 'active touch'.

During his years at Liverpool, Sherrington discovered the afferent nerve fibres running from the muscle-spindles into the dorsal roots; helped the then-unknown Ruffini to publish his classical description of the muscle-spindle and Golgi tendon organ;[10] and wrote his seminal chapter on 'The Muscular Sense' in Schäfer's great *Textbook of Physiology* (1900). In his adjoining chapter on 'Cutaneous Sensations' he was the first to define and differentiate 'active' and 'passive touch'. He was specially interested in the effects on movement of depriving the monkey's central nervous system of afferent impulses from one arm by section of the dorsal roots. The useless arm 'curiously closely simulates the impairment of motility ensuing on ablation of the limb region of the cortex cerebri', suggesting 'that not only the cortex, but the whole sensory path from the periphery to the cortex cerebri, is in action during voluntary movement.'[11]

Second, experiments on the cortical neuronal hardware that receives and processes information from the hand, and on the corticospinal neurones that most directly control the elemental movements of the fingers.

Sherrington was very interested in 'the excitable cortex' and its projection to the spinal cord. His interest was aroused in 1881 when he assisted Langley in the anatomical description of the right half of the brain of Goltz's nearly-decorticate dog. This dog, with its remarkable degree of restitution of motor behaviour, had been exhibited at the International Medical Congress in London at Goltz's historic confrontation with Ferrier.[12] Ferrier's hemiplegic monkeys led him to affirm the doctrine of cerebral localization. Goltz's decorticate dogs led him to deny it. In the 1880s there was much inconclusive controversy about the 'motor', 'sensory' or 'sensorimotor' nature of cortical areas. Wishing to distance himself from these controversies, Sherrington proposed the term 'cord area'[13] to denote 'all this region of the cortex cerebri which is so directly connected with the spinal cord that lesions in the former induce degeneration in the latter.' The term was never accepted. This cortico-spinal interest culminated in his experiments with Grünbaum (Leyton) on electrical stimulation of the 'excitable cortex' of the great apes.[14]

CHAPTER II

Categorization and investigation of movements

Some consideration of general strategies of research on movements will serve as prelude to the two leading themes.

Postures and movements are the products of the nervous and musculoskeletal systems. It seemed sensible to start from the product and then work back to the neural hardware that produces it. In so doing, the Lecture largely ignored the details of the musculoskeletal contribution, in spite of the important and exciting research that has been and is being done on the biophysics of muscle and the biomechanics of muscle and joint. The musculoskeletal apparatus was regarded simply as a transformer of patterns of nerve impulses into elegantly-coordinated movements of appropriate speed and power.

1. *Categorization of movements*

The Victorian physician-philosopher J. Hughlings Jackson[1] graded movements on a scale ranging from 'most automatic' to 'least automatic'. A few illustrations will show that this categorization is still useful. The simplest segmental reflexes and automatisms belong to the lowest grade on the scale, skilled voluntary movements to the highest. In the competition for entry to the Final Common Path to the musculoskeletal apparatus, the 'most automatic' nociceptive flexion-withdrawal reflex takes priority over less urgent reflexes, e.g. scratching. 'Less automatic' movements of the limbs include those of locomotion, which are driven by a central neuronal 'pattern generator' whose activity can be modified by inputs from the limbs as well as by signals descending from 'least automatic' levels of the brain. These signals enable the animal to control its speed, to avoid obstacles, to move towards some situations and to escape from others. Use of the hand in arboreal locomotion is 'more automatic' than its use in tactile exploration and prehension. 'Less automatic' functions are differentially vulnerable to lesions of the forebrain and its projection pathways. A monkey can cling, climb or clutch for support with a hand which has been rendered useless for tactile exploration, manipulation and skilled performance in general. A patient who following a stroke can no longer easily button his clothes, may still be able to lift a chair. The 'more' and 'less automatic' movements of the eyes have already been compared to those of the hand.

The respiratory movements are highly automatic. Their central 'pattern generator' goes on producing movements even under general anaesthesia deep enough to eliminate all other movements, all posture, and most spinal reflexes. The respiratory movements do not, like those of the hand, act mechanically on solid objects in the external environment: their biological goal is to satisfy the fluctuating demand of the internal environment for oxygen. Their power is increased by servo-action in response to increased airway resistance. But the same musculoskeletal apparatus also shares responsibility for the posture of the trunk, and hence for the stability of the shoulder-girdle ('less automatic'), and for speech, song and the playing of wind instruments ('least automatic'). In the competition for entry to the Final Common Path, the 'more automatic' functions are somehow continuously fulfilled, whatever 'priority interruptions' may be made by the 'least automatic'.

In man, skills which have been trained and perfected by long practice, so that they can be run off with little or no conscious attention, have become 'more automatic' than the first careful approaches to an unfamiliar skill.

2. *From 'most automatic' movements to neurone and synapse*

Sherrington's general strategy at Liverpool was to start with reflex movements—relatively simple, highly stereotyped automatic movements of scratching, stepping, flexion-withdrawal and crossed extension, elicited from what he later[2] described as an actual Cartesian puppet, the spinal or decerebrate mammal. Such items of behaviour have been recognized, since the work of Marshall Hall in the 1830s, as dependent on inherited structure in the nervous system and not on learning by the individual.

In Sherrington's earliest experiments it was the movements themselves, and their biological goals, that occupied the centre of interest. In removing a flea from the skin, the scratching limb was moved by the coordinated actions of nineteen muscles, supported posturally by seventeen more. In suddenly withdrawing a foot from a thorn, the limb girdle was saved from collapse by extension of the opposite leg. These performances were interesting because it could be assumed that their essential cellular basis was the same as that of the less stereotyped, less automatic movements which could not have been explained, in those days, in terms of elemental neuronal properties. The reflexes were examined first in intact limbs by hand and eye, then by crude recording of the excursions of an isotonic lever attached by a thread to the intact limb. Designed and interpreted by Sherrington's genius, these experiments established that the essential properties of reflex action, and, by extension, of all central nervous action, depend ultimately on the intercellular contacts which he named synapses.

As the years went by, his interests moved away from reflexes as items of animal behaviour and towards the fundamental phenomena of central synaptic excitation and inhibition, conceived as summing algebraically at the motoneurone membrane. This first opening of a 'black box' in the CNS was made possible by quantitative high-speed isometric myography of a single muscle, and of pairs of muscles acting antagonistically at a joint, combined with string-galvanometric electromyography. The tendons were detached from their insertions and the muscles were isolated except at their bony origins, with preservation of their nerve- and blood-supplies. The muscles served only as measurers of the firing of motoneurones and of the inhibitory suppression of their firing. Other muscles whose contractions would have been coordinated with the isolated muscles in natural movements were paralysed by cutting their nerves or tendons. This was necessary to avoid any mechanical or electrical interference with the delicate instrumental recording from the output detector muscles. These experiments[3] gave quantitative precision to the qualitative inferences from the earlier work.

3. *From 'least automatic' movements to hypothetical neuronal networks*

Sherrington's strategy was to start from 'most automatic' movements and work towards their cellular basis. A different strategy was adopted by N. Bernstein and his pupils in Moscow between and after the two World Wars. They started from skilled 'least automatic' performances such as hammering a nail, using a file, and striking piano keys, and resolved their four dimensional structures by ingenious methods of the greatest accuracy and refinement.[4] They were working towards complex neuronal networks rather than short chains of cells. The analysis of these beautiful records—cyclograms, kymocyclograms and mirror-kymocyclograms—led Bernstein to some important predictions and generalizations about how the brain controls movements (Fig. 2). These inferences have the status of 'black boxes', awaiting attachment to particular neuronal hardware—in forebrain, cerebellum, brainstem, cranial and spinal segments—and awaiting opening whenever appropriate techniques become available. But before there was any opening of 'black boxes', and from the analysis of movements alone, Bernstein was able to set limits to the ways in which the cerebral cortex could plausibly operate in producing skilled movement, and to criticize theories based on the results of electrical stimulation of the motor areas in monkeys and men.

Cyclography, 'the planar projection of movements', was derived indirectly from Marey's classical methods. Cyclograms (chronograms, chronophotographs) are illustrated in Figs 1 and 3. Point sources of light were attached to appropriate points on the limbs, and their two-dimensional trajectories photographed on a stationary plate. To incorporate the time dimension the camera was fitted with an accurately-calibrated rotating shutter. In the resulting photographs the two-dimensional trajectories appeared as a series of dots, which remained separate at frequencies up to 600 Hz; the intervals between the dots revealed the speedings-up, steady velocities and slowings-down of the movements.

To obtain non-overlapping records of repeated movements Bernstein invented kymocyclography, in which the stationary plate was replaced by a continuously-moving film. The separated movements were subsequently reconstructed by subtracting the movement of the film. To obtain stereoscopic records he invented mirror-kymocyclography, in which a single camera received, side-by-side, a direct and a reflected image. The position of the mirror was so arranged as to be equivalent to photography with two cameras at an appropriate separation. Thus only one rotating shutter was required, avoiding the need to synchronize the shutters of two cameras.

These methods measure only the kinetics of movement. To measure the dynamics one needs to know the masses as well as the accelerations of the moving parts, and Bernstein duly invented methods of estimating these masses in living men and women.

4. *Programmes and control circuits for movements*

'Photographic registration of movements is in no way the final aim of investigation but merely its raw data,' wrote Bernstein. The real

Fig. 2 Bernstein's 'simplest possible block diagram of an apparatus for the control of movements'. See text. (Bernstein, *The Co-ordination and Regulation of Movements*, 1967, Oxford: Pergamon Press)

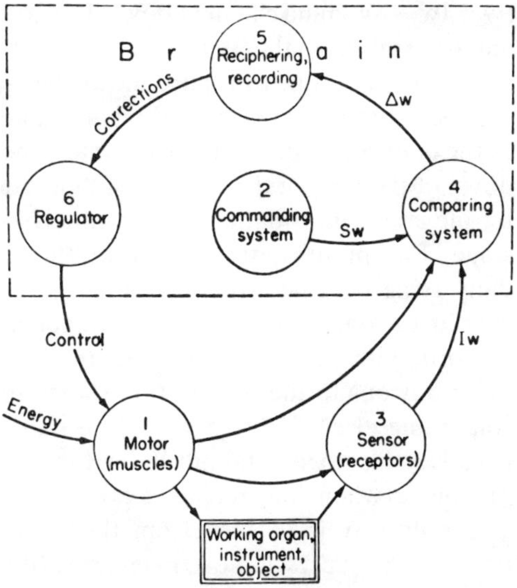

goal is 'physiological and biomechanical analysis of the processes of movement.' Analysis of such actions as hammering, filing and striking piano keys led him to the conclusion that these were not merely 'arbitrary collections of successive reflex elements,' but were marked by what he variously called 'integrity', 'structural completeness', 'structural physiology' and specific four dimensional 'morphology'. 'A movement never responds to detailed changes by a change in its detail; it responds as a whole to changes in each small part.'

His formulation of the general principles of control of movement includes the notion of hierarchical levels of control: first, the overriding engram, 'motor problem', 'motor image' or 'programme of a movement' (Bewegungsformel, Bewegungsgestalt), organized in advance and *simultaneously* present *in toto*; next, the subordinate engrams of the 'operational stages' in which appropriate combinations of muscles are activated *sequentially* in the 'ecphoration' of the movement. Fig. 2 reproduces his 'simplest possible block diagram of an apparatus for the control of movements' (1957). The 'commanding system' in the brain (2) specifies the *required* time-course (Sw, Sollwert) of the motor output (1). Proprioceptive signals from the muscles and exteroceptive signals from the manipulandum (3) detect the *actual* time-course (Iw, Istwert) of (1). Back in the brain a 'comparing system' (4) measures the magnitude of mismatch between Sw and Iw (w = Iw − Sw), and the next box (5) re-codes the necessary correction and transmits it to the 'regulator' (6), the brain's output to (1).

Self-observations of a pure motor hemiplegia by a neurologically-sophisticated patient[5] showed what happens when the 'ecphoration' of the overriding 'programme of a movement' is frustrated by a breakdown of the temporal sequence of 'operational stages', each set going by incoming information of the timely completion of the previous stage. Two months after the stroke this patient began to resume the daily tying of a bow tie. Seven to ten attempts were necessary on each occasion. 'The appropriate finger movements were difficult to perform with sufficient strength, speed and coordination, but it was quite obvious to the patient that the main reason for the failure was something else ... Subjectively the patient felt as if he had to stop because "his fingers did not know the next move." He had the same feeling as when one recites a poem or sings a song and gets lost. The only way is to start from the beginning. It was felt as if the delay in the succession of movements (due to pareses and spasticity) interrupted a chain of more or less automatic movements. Consciously directing attention to the finger movements did not improve the performance; on the contrary it made it quite impossible.' The forty-year-old 'overriding programme', perhaps, was still intact, but the motor cortical output was damaged. In the absence of a normal succession of completed movements, each contributing its quota of feedback within a critical period of time, the programme somehow 'got lost' and failed to retain control of the performance. Fortunately this patient has made an excellent recovery.

Bernstein's theoretical (black-box) 'regulator' (Fig. 2(6)) can be attached, in part at least, to the real cortical motor area. Electrical stimulation of this area of the human cortex in the nineteen-twenties[6] evoked a wide range of more-or-less circumscribed motor responses from 'points' which gave repeatable responses, provided that—as Leyton and Sherrington had found necessary in apes—a sufficient interval was allowed between successive stimulations. The term 'punctate localization' came into general use, and it would have been generally assumed that in natural movements, as Bernstein put it, 'the central impulse *a* always produces movement *A*, and impulse *b* always produces movement *B*, from which it is easy to proceed to a representation of the motor area of the cortex as a distribution panel with push-buttons.' But Bernstein's mathematical treatment showed that the same 'impulse' 'may produce completely different effects because of the interplay of external forces and because of variations in the initial conditions . . . a determinate effect is possible for a movement only in a case where the central impulse is very different under different conditions, being a function of the positions and the velocities of the limbs.' He concluded that 'the push-button type of one-to-one correspondence does not exist and the cerebral motor area organizes responses by deftly adjusting and balancing between resultant external forces and the manifestations of inertia, constantly reacting to proprioceptive signals and simultaneously integrating impulses from separate central subsystems, so that ten successive repetitions of the same movement demand ten successive impulses all different from each other; and the presence in the cortex of localizational equipment of the Hitzig or Foerster type begins to seem a very dubious interpretation.'

More will be said later (Chapter IX) about the 'localizational equipment' of the cortex in the light of subsequent research. With regard to the higher levels of function, Bernstein's hypothesis, as far back as 1935, was that they contain 'projections of space and not projections of joints and muscles.' Whether written small or large, on a horizontal paper or on a vertical blackboard, an individual's handwriting remains recognizably his own.

5. *Ballistic and ramp movements*[7]

Movements in general, whether more or less automatic, can be categorized in yet another way. *Ballistic* movements are brief, starting and ending abruptly; *ramp* movements start gradually, continue more or less smoothly, and are more or less prolonged.

As classically defined, a ballistic movement is a literal *throwing* of a limb or part of a limb by a brief muscular contraction. The knee-jerk is an ideal example of such a movement—in this case, a reflex movement—in which the muscle relaxes before the movement has been completed. But the definition has been widened to include, and even to give first place to, another idea: that a voluntary ballistic movement, once launched, completes its brief pre-programmed trajectory before it can be influenced by any feedback from the moving parts, or from vision. This is known in current jargon as 'open loop' operation. Such a movement might or might not be ballistic in the classical sense that the muscle had relaxed before the movement was completed. In the striking of piano or typewriter keys, for example, the new emphasis would dwell on the pre-programming of the required trajectories, forces and speeds by the brain, on the basis of the initial positions and masses of the moving parts; and on the completion of the muscular *contraction* in so short a time that no proprioceptive or visual feedback could have any influence on the *movement*.

Such preprogrammed movements are characterized electromyographically by a stereotyped sequence of discharges. The antagonist muscle (if it was discharging) is first silenced. There is then a brief burst from the prime mover, followed by a brief burst from the antagonist, and ending with a second burst from the prime mover (see Fig. 4).

In contrast, a ramp movement begins smoothly, and proceeds at a more or less steady speed (see Fig. 8B). From the moment at which it has been going for long enough to allow their signals to reach the central nervous system, it becomes continuously responsive (though with a delay) to proprioceptive and visual feedback ('closed loop' operation).

Ballistic and ramp movements are convenient experimental abstractions from the complex totality of movement, and much is being learned from them about our everyday movements which are, of course, often compounded of both elements.

CHAPTER III

Programme and feedback in simple voluntary movements

It is time to descend from such 'least automatic' performances as hammering a nail or tying a bow tie to the apparently simple everyday act of reaching for and grasping an object in the field of vision.[1]

Fig. 3 shows experimental records of this act.[2] The upper record A is a chronophotograph of a movement that was completed in about one second. In it two components can be distinguished: a transportation component that brings the hand to the target along a curved trajectory about 50 cm in length, and a manipulation component that shapes the hand in preparation for the grasp. The opening of the hand begins at the outset of the movement: 'il faut remarquer la précocité et la finesse avec lesquelles la main se prédispose à la saisie de l'objet'.

1. *Programme and visual feedback*

With Bernstein's diagram in mind (Fig. 2), we should be asking questions about the relationships between the programme of the movement (Sollwert) and the incoming signals (Istwert) generated as the movement approaches its target.

It is obvious that the initial positions of the trunk, arm and hand must somehow get incorporated into the programme. In a blindfolded subject these could be specified by proprioceptive signals, or by visual storage ('memory') of their initial pre-blindfolding positions, complemented by proprioceptive signals that these positions had not changed.

The reaction-time of the hand to a visual signal is about 300 ms, by which time the movement we are considering would already have completed about one-third of its total trajectory. Up to this point, therefore, it must be wholly preprogrammed with respect to the approach to its visual goal—'ballistic', in the sense that the trajectory of a thrown ball passes out of the thrower's control as soon as it leaves the hand. Thereafter, visual information of the progress of the movement becomes continuously available, but lagging always about 300 ms behind the moving front.

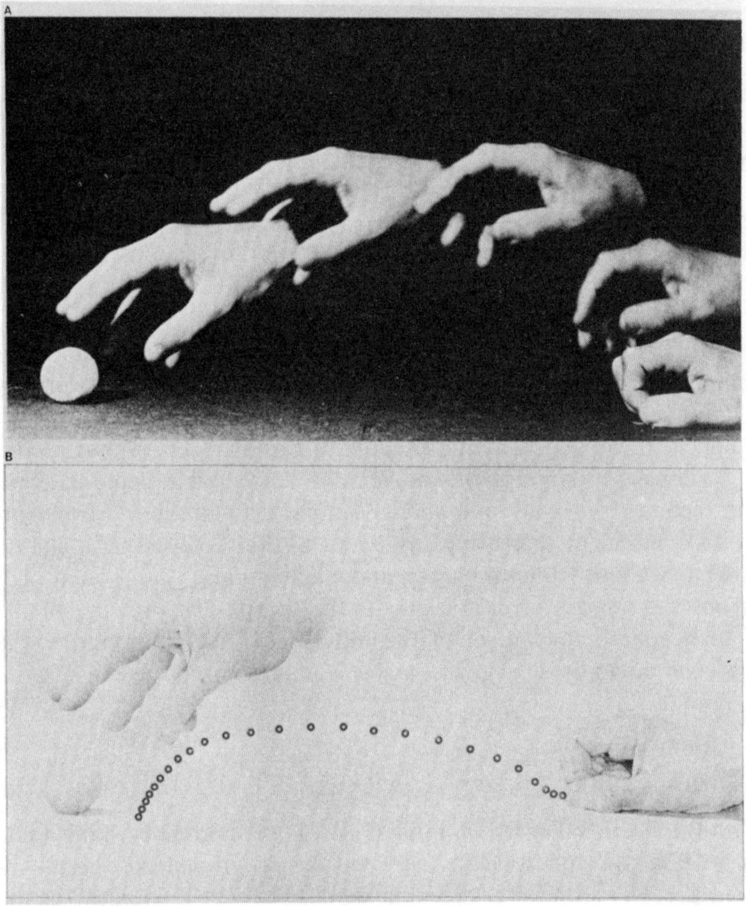

Fig. 3 The act of reaching and grasping. A, chronophotograph; B, trajectory of marker attached to tip of thumb, plotted from successive frames of ciné film taken at 25 frames per second. See text. (Jeannerod, *La Recherche* **12,** 376–78, 1981)

The next experiment shows what happens when vision of the movement is excluded, the target alone remaining in view. The same movement as in Fig. 3A was performed in the lower compartment of a box divided into upper and lower compartments by a semi-reflecting mirror. Looking through a window in the upper compartment, the subject could see his forearm, hand and target in the lower compartment. The trajectory of the tip of the thumb was plotted from the frames of a ciné film taken at 25 frames S^{-1} (Fig. 3B). To exclude vision of the moving arm, but retain that of the target, a mask was placed below the mirror, making the contents of the lower compartment invisible from the window in the upper compartment; a second target was so located in the roof of the upper compartment that its reflection (virtual image) appeared to coincide exactly, in size and position, with the real target in the lower compartment. The experiment showed that the trajectory of the blind movement is perfectly formed, its acceleration and deceleration perfectly controlled, but that the movement always falls a few cm short of the target (Fig. 3B). So, when the final two-thirds of the movement are deprived of the visual signals that were available in the experiment of Fig. 3A, the movement does not disintegrate. It remains a perfect pre-programmed movement. But, under these experimental conditions at least, vision is essential for the accuracy of the final 'homing-in' on the object that is to be picked up.

Nothing has been said about the act of reaching and grasping by the blind. Would the initial exploratory location of the target set up a proprioceptive transportation-and-manipulation programme which would operate accurately first time round? Or would the first trial fall short of the target; if so, how many trials would be needed to eliminate the error?

2. *Effects of loss of proprioceptive feedback from the moving parts*

We cannot of course assume that the movement produced in the absence of vision (in the 'ballistic' or 'visual open-loop' mode) is being controlled by the programme alone, and that it can do so without proprioceptive information from the moving parts—that is, without the flow of information represented by Bernstein's arrows $1 \rightarrow 3$, $1 \rightarrow 4$ and $3 \rightarrow 4$ (Fig. 2). Evidence to be given below (p. 37) shows that proprioceptive feedback could begin to affect the trajectory and velocity of the movement, and to correct any deviations from the programme, about 40 ms after its onset. So the first twenty-fifth only of the movement, and not the first one-third, need be in a wholly ballistic mode. From that point onward the motor output could be, as Bernstein supposed, 'constantly reacting to proprioceptive signals', though these would be lagging always 40 ms behind the front of the movement. It seems highly unlikely *a priori* that the perfectly coordinated, visually-preprogrammed movement illustrated in Fig. 3 could be performed at all, whether in the visual-open-loop or visual-closed-loop mode, if all proprioceptive feedback could be suddenly withdrawn.

Unfortunately the flow of proprioceptive feedback cannot be easily and reversibly interrupted, as can the flow of visual feedback by merely sliding a mask under a semi-reflecting mirror (as in Fig. 3B). One cannot simply compare performances in the visual open-loop mode in alternate trials, first with, then without, proprioceptive feedback. Since this experiment is impracticable, the importance of feedback has to be pieced together from the results of some less-than-ideal observations and experiments, each, however, making its contribution to the total picture.

The importance of feedback was discovered in monkeys by cutting the dorsal roots.[3] This experiment, however, cuts off *all* feedback from the moving parts which enters the CNS by the dorsal roots: not only that from muscles and joints (proprioceptive), but also from skin (exteroceptive). The term response feedback includes both of these. A further difficulty is that the deafferentation may not even be complete. The validity of the Bell–Magendie Law—that the dorsal roots are purely sensory and the ventral roots purely motor—has been challenged by experiments showing that the ventral roots contain many unmyelinated axons which enter them 'aberrantly' from the dorsal root ganglia.[4] So dorsal-root ganglionectomy, and not just dorsal rhizotomy, would be necessary to be sure of complete deafferentation, and its completeness would have to be verified histologically *post-mortem*.

But if acute deafferentation by dorsal rhizotomy, being irreversible as well as unselective, falls short of the ideal experiment, it does reveal the devastating 'paralytic' effect of a sudden massive withdrawal of response feedback. In the months following the operation, however, it is possible to encourage some recovery of motor function. How such compensation is brought about remains a problem: how the brain can adapt to the loss of response feedback, and develop new programmes to compensate in some degree for the loss.

Studies of patients with chronic pathological processes involving the central somatic afferent pathways, for example, small plaques of multiple sclerosis, or peripheral sensory neuropathies without significant involvement of motor axons, have much to contribute. Their disadvantage is that their onset is apt to be insidious rather than acute, so that, by the time the patients come to neurophysiological investigation, compensatory changes may be well under way.

3. *Surgical deafferentation*

Sherrington summarized his experience on the acute disability following unilateral dorsal root section thus: 'Willed movements are upset in the extreme. In the monkey the desensitised limb is practically useless, for prehension quite so. The animal soon treats it as useless and as an encumbrance worse than useless; it attacks it and would tear it away. . . . Wild incoordination under willed action; little or no abnormality under the action of the directly stimulated motor cortex.'[5]

Later research has revised Sherrington's account of the degree of disability.[6] So long as injury to microscopical blood vessels running with the dorsal root filaments is avoided, so that the spinal cord suffers no ischaemic damage, and so long as the normal arm is restrained, or both arms are deafferented, a monkey will try to reach for grapes within a few hours of recovery from the general anaesthetic. Its dependence on vision is obvious. 'The animal attends closely to his arm and spends much time observing attempts to move the arm. Moreover, we have not seen reaching movements soon after surgery unless the offered object and the hand were in simultaneous view.' Thus the immediate effect is to make a programmed reaching-and-grasping movement impossible, even in the visual-closed-loop mode.

The first successful grasping occurred after two weeks' daily training, during which help had to be given in forming a grasp. But though a monkey eventually learns to reach reliably for food, its movements remain permanently ataxic and dysmetric. 'Although movement is not abolished and single components of a forelimb repertoire have been trained by various investigators, the elegance of normal movements is lost.'

The compensatory processes which enable a deafferented monkey to learn to perform a limited repertory of clumsy movements remain elusive, but invite speculation. Suppose that a visually-programmed performance of the deafferented hand includes, as an inseparable component, some associated posture or movement in a remote area of the body, say trunk or legs, whose afferent innervation is intact. The movement of the deafferented arm shifts the body's centre of gravity and elicits postural responses in the trunk and legs. Then mutual reinforcement between visual feedback from the hand and response feedback from the trunk and legs could conceivably play an important part in re-learning to reach and grasp. Monkeys are 'quite capable of such subterfuges and suitable controls for them have not been incorporated into some of the behavioural tasks which have been used.'

Another factor that could conceivably play a part in compensation is 'internal feedback', also known as 'efference copy' or 'corollary discharge'. Cortico-spinal axons send collateral branches into the somatic afferent pathways ascending from the spinal cord to the brain. These collaterals form synapses with first-, second- and higher-order neurones in the 'relay' nuclei of the ascending pathways, and supply the brain and cerebellum with 'copies' of the 'commands' addressed by the motor cortex to the spinal cord in the 'ecphoration' of the pre-programmed sequences of movement. Deafferentation would have deprived the 'relay' neurones of their somatic afferent synapses as the result of degeneration and disuse. The vacated post-synaptic sites could then have been invaded by new synaptic boutons formed by the sprouting of corticospinal axon collaterals, with a corresponding enhancement of their actions on the 'relay' cells. It is conceivable that mutual reinforcement between visual feedback from the hand and a much-intensified 'efference copy' generated by motor commands addressed to the hand, could be an important factor in re-learning to reach and grasp.

Neurosurgeons[7] have sometimes cut the cervical dorsal roots in patients in order to relieve spasticity by abolishing their tonic stretch reflexes. The results are inescapably complicated by the brain damage that was causing the spasticity they sought to relieve. Even in the acute stage, however, and even when blindfolded, the patients could still make elemental movements. They could flex, extend, raise, lower and rotate their arms and pronate, supinate, extend and flex their hands. The start of these movements was delayed, but they were prompter and more accurate if vision was allowed. Movements of the fingers were worst affected, whether under visual guidance or not. When the patient was asked to move a named finger the wrong finger would respond, or if the correct finger were moved it would be accompanied by other fingers or by movement of the whole hand.

4. *Deafferentation by spinal lesions*

A second source of evidence comes from patients suffering from interruption of central somatic afferent pathways. In the days when tabes dorsalis was very common, 'brachial tabes' was rare. 'The commencing defect is revealed by delicate movements, such as writing. As it increases, all movements become irregular; it is impossible for the patient to button his coat, or to pick up a small object from the table; the fingers twist about in the attempt. The grasp is not sustained; first one finger is felt to relax and then another . . . Even with extreme incoordination, the power of the muscles may be unimpaired.'[8] Similar mischief can be caused by a plaque of multiple sclerosis at the top end of one of the dorsal columns.

5. *Deafferentation by peripheral sensory neuropathy*

A third, very rich, and relatively uncomplicated source of evidence can be provided by patients suffering from distal peripheral neuropathies with negligible affection of motor axons. One such case has been examined in great detail, and is a model for future studies.[9] The diagnosis was based not only on expert clinical examination but also on objective electrophysiological measurements. The afferent axons of the median nerves were shown to be blocked between wrists and elbows; and electric shocks to the fingers evoked no electric waves at the top end of the spinal cord or in the sensorimotor cortex. The motor axons conducted impulses at normal velocity, and electromyography showed no denervation potentials; there was, however, slight bilateral weakness of the intrinsic hand muscles. Before these objective methods became available there might just have been room for some inconclusive discussion as to whether the symptoms were organic or hysterical. They started insidiously, and had been present for about two years at the time of investigation. The patient's ability to make preprogrammed elemental movements is the only question at this point: his more elaborate performances, and the reflex and sensory phenomena, will be taken up later.

Without practice this patient, even when blindfolded, was able to oppose the tip of the thumb to the tip of any named finger of the same hand. This finding conflicts with the neurosurgical experience that the wrong finger, or more than one finger, was moved. Possible explanations of the difference between the neurosurgical and neuropathological deafferentation are the absence of organic brain damage, and of possible operative injury to the spinal cord; and the confinement of deafferentation to structures distal to the elbow, in contrast to the neurosurgical deafferentation of the whole limb and of its attachment to the trunk. The more extensive deafferentation would have deprived the subject of possible cues about the fingers arising from the more proximal segments of the limb, which included the origins of the long flexors and extensors of the fingers.

Again without practice, the patient could outline with the tip of an index finger in the air in front of him a square, a circle or a figure-of-eight. These near-normal performances were recorded by cyclogrammetry. But though these, and the independent movements of the digits, could be initiated in the absence of vision, they could not be sustained. Without visual feedback they disintegrated progressively over the course of about 30 s. So the flow of somatic afferent information is necessary if the selected programme is to continue to emit the appropriate sequences and combinations of muscular contractions and relaxations. To be able to start the performance when blindfolded depends presumably on visual storage ('memory') of the initial positions; the progressive disintegration would depend on a progressive drift of the preprogrammed outflow as the short-term visual storage fades, and no updating (Istwert) is forthcoming from the moving parts.

In contrast with the surprisingly minor degree of disorder revealed by the experimental tests, the disability in the much less automatic peformances of everyday life was severe. To this we shall return.

6. *Partial deafferentation of the motor cortex*

Very rarely it will be possible to find a patient with a lesion confined to the somaesthetic receiving and association areas of the cerebral cortex, without damage to the precentral motor areas. Modern imaging techniques can localize such a lesion with fair confidence, and anticipate to some extent the findings of an eventual neuroanatomical autopsy. Such a case would have the advantage that the exteroceptive and proprioceptive inputs to lower levels of the central nervous system would remain intact, the motor area being more or less selectively deafferented, at least with regard to its cortico-cortical input from somaesthetic areas: its thalamocortical input (Chapter IX) could however be more or less preserved. It is unlikely that neuropsychological and neurophysiological investigation would be possible in the early stages following the onset of such a lesion, and indeed, in the one case so far reported, the full investigation was delayed for a number of years.

The visually directed act of reaching and grasping which introduced this chapter was investigated in this single case.[10] 'Without vision, actions involving the fingers could not be sustained, and initiation was difficult if not impossible . . . The reaching phase of the movement was correctly executed, although the grasping phase was misadapted due to the inability to shape the hand according to the spatial configuration of the object.' It can be assumed that if *acute* withdrawal of proprioceptive feedback were practicable in man, all parts of the act of reaching and grasping would become impossible immediately.

CHAPTER IV

Control of an elemental voluntary movement

To the neuropsychologist, the acts of pointing (Fig. 1), reaching and grasping (Fig. 3), and outlining geometrical shapes in the air are among the simplest accessible to study. To the human neurophysiologist the simplest act is flexion or extension of a single manual hinge joint that has no other degrees of freedom. He wishes to come as near as possible to Sherringtonian myography and electromyography of a single muscle or an antagonistic pair. The interphalangeal joint of the thumb is especially suitable. No simpler brain programme can be imagined than that which produces its simple angular trajectory, variable only in range, isotonic and isometric force and angular velocity.[1]

The recording system includes a movement transducer (goniometer), coaxial with the joint, which displays the angular movements on the screen of a cathode-ray tube, and with them, markers showing the starting-points and targets of the demanded movements. The displays can be blanked off whenever the subject is required to perform the movements without visual feedback.

Movements of the interphalangeal joint of the thumb are effected by the two extrinsic muscles, flexor pollicis longus (FPL) and extensor pollicis longus (EPL). The movements and the rectified electromyograms of these antagonists are shown in Fig. 4. Clamping the proximal phalanx and splinting the hand prevents the intrinsic muscles from interfering with the recorded movements (though they may still exert isometric force at the carpometacarpal and metacarpophalangeal joints). The time course of the movement is recorded on tape or film, and also the derived time-course of its velocity (Fig. 4). The goniometer lever that is moved by the thumb incorporates a force transducer for measuring isometric force. Also coaxial with the joint is a special torque motor which can be set to provide elastic or viscous resistance to the movements, and can unload, halt, or forcibly reverse them when the experimenter wishes to inject such perturbations in order to test and time the responses of the muscles' control loops (see below, p. 37).

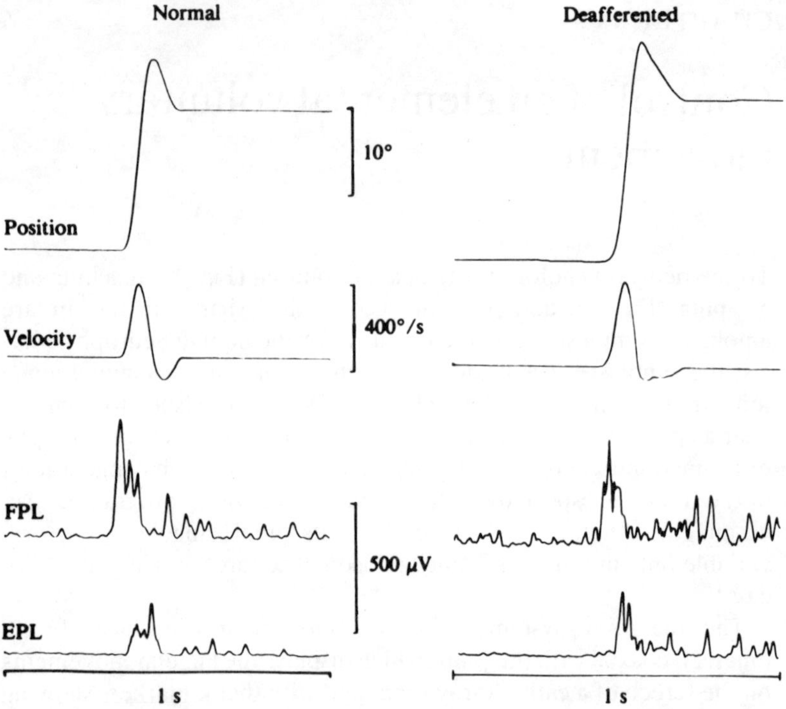

Fig. 4 Ballistic isotonic flexion at interphalangeal joint of thumb, by normal person and a patient whose hands were deafferented by a peripheral sensory neuropathy. *Upper records:* angular displacement and velocity; *lower records:* rectified electromyograms from the prime mover (FPL) and the antagonist (EPL). Time marker 1 second. Averaged records of 16 trials. See text. (Rothwell *et al.*, *Brain* **105,** 515–42, 1982)

The relative contributions of programme and response feedback to the control of an elemental movement are clearly brought out by comparing the performances of normal subjects with those of the patient whose forearms and hands were deafferented by peripheral sensory neuropathy.[2]

1. *Elemental ballistic movements*

Fig. 4 compares the mechanical events in flexion of the interphalangeal joint of the thumb, and the electrical responses of the agonist muscle FPL and its antagonist EPL, in a healthy subject and in the deafferented patient. Both were asked to flex the joint as rapidly as possible. In emitting this brief ballistic movement the patient's preprogrammed

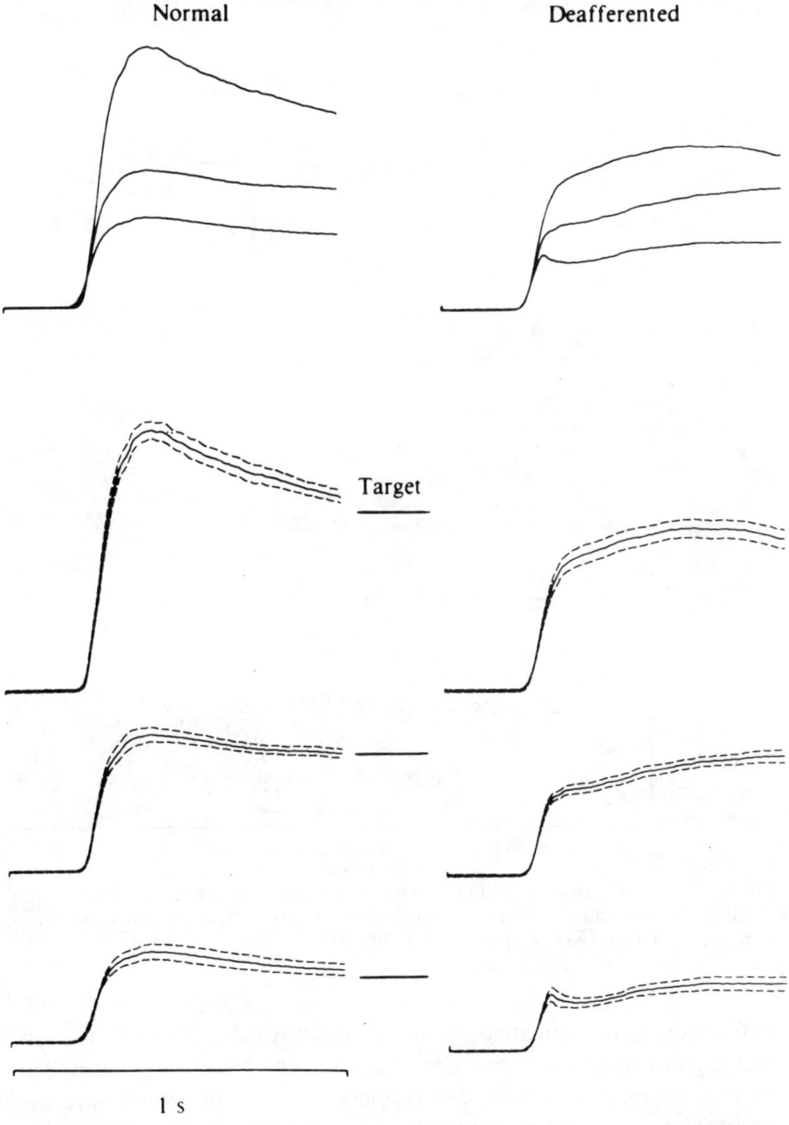

Fig. 5 Ballistic isometric flexor torque exerted against stationary lever by distal phalanx of thumb by normal person and deafferented patient, aiming at three target force levels learned initially with a visual display and then performed in the absence of vision. Time marker 1 second. Averaged records of 16 trials. *Full lines*, means; *dotted lines*, standard errors. See text. (Rothwell *et al.*, *Brain* **105,** 515–42, 1982)

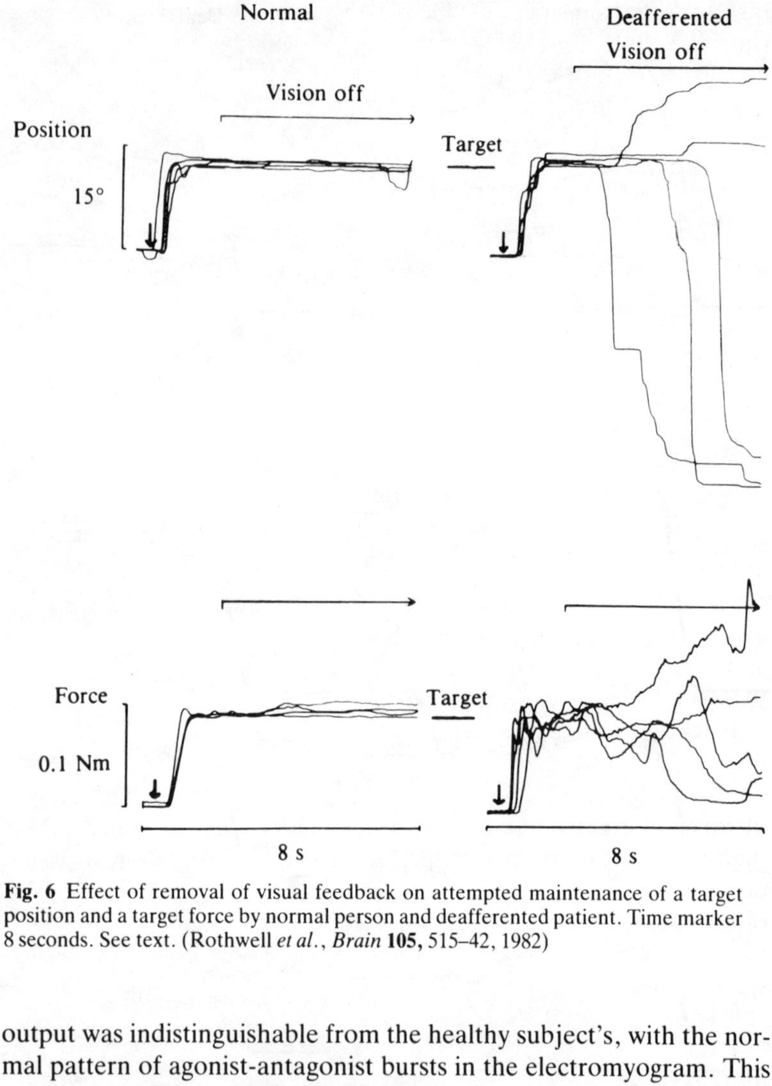

Fig. 6 Effect of removal of visual feedback on attempted maintenance of a target position and a target force by normal person and deafferented patient. Time marker 8 seconds. See text. (Rothwell *et al.*, *Brain* **105,** 515–42, 1982)

output was indistinguishable from the healthy subject's, with the normal pattern of agonist-antagonist bursts in the electromyogram. This pattern is therefore wholly determined by the brain-programme, and is independent of proprioceptive feedback. Indeed, for a short period during and following the agonist burst, a proprioceptive afferent volley is actually blocked centrally, and fails to evoke its usual 'long-loop reflex' (see below, p. 37).[3]

After some visually-guided training in the 'thumb machine,' the patient could also learn to flex the thumb rapidly to any of three target positions, completing the movement in about 0.1 s. The records were indistinguishable from the normal whether the visual display was blanked off or remained present.

Another task was to generate, as rapidly as he could, one of three prescribed levels of isometric force (Fig. 5). The patient trained himself by watching the visual displays of the target force levels as well as those of the output of the force transducer he was pressing with his thumb. He could then achieve the target forces accurately when the displays were blanked off.[4]

2. *Elemental ramp movements*

The frontiers between ballistic and ramp movements, and between independence-of and dependence-on proprioceptive feedback, are not sharp. When the patient, having learned to flex his thumb ballistically to any of three target positions, trained himself to make the same movements five to ten times more slowly, he could still accurately attain the three target positions without vision, but the trajectories of the movements became more variable. Presumably the effects of loss of proprioceptive feedback were beginning to declare themselves as the detailed control of the trajectory by the programme began to break down with lapse of time.

When, however, the task was to maintain a target position or target force for longer than about 0.5 s after the visual display was blanked off, the control subjects succeeded, but the patient failed: his outputs of force or joint angle began to fluctuate wildly on either side of their targets (Fig. 6). Since these tasks can be performed by people deprived of inputs from skin and joint by blocking the digital nerves of the thumb, the patient's failure would have been due specifically to loss of information from his intramuscular receptors, the muscle spindles and tendon organs.

3. *Responses of intramuscular receptors to passive and active movement*

Before going further it may be helpful to offer the sketchiest possible *aide-mémoire* of the properties of the muscle spindles and Golgi tendon organs—properties that enable them to measure the length, rate-of-change of length and force developed by a muscle, and to signal these measurements to the central nervous system.[5]

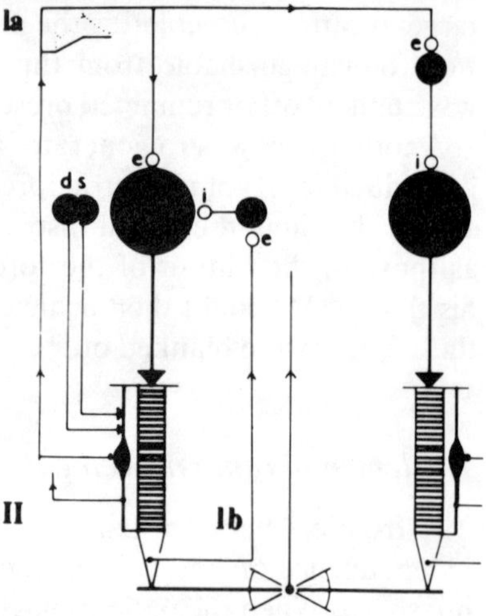

Fig. 7 Mechanical and neuronal coupling between a pair of muscles acting at a hinge joint. *Left side of diagram* represents flexor pollicis longus when it is acting as a prime mover; *right side* represents its antagonist, EPL. To keep the diagram as uncluttered as possible, the symmetrical proprioceptive networks of EPL, and the Ia-inhibitory pathway from EPL to FPL, have been omitted. The muscle spindles are correctly represented as coupled 'in parallel' with the skeleto-motor muscle fibres, but they do not, as in the diagram, run the full length of the muscle: they are only 1–3 mm in length and are scattered throughout the substance of the muscle. *e*, excitatory synapses; *i*, inhibitory synapses. See text

Fig. 7 presents a crude dogmatic diagram of a pair of muscles that act antagonistically at a hinge joint—as it might be, flexor and extensor pollicis longus acting across the interphalangeal joint of the thumb. The joint is represented as a see-saw, as a reminder that neither muscle can contract without lengthening the other. The lengths of the two muscles, therefore, correspond to particular angles at the joint.[6] The diagram illustrates the mechanical couplings between the different classes of receptors and the moving parts.

The muscle spindles incorporate afferent nerve-endings of two types, 'primary' (Fig. 7, Ia) and 'secondary' (Fig. 7, II). These are stimulated by lengthening the muscle, as by contraction of its antagonist, or by passive movement of the joint in the course of psychophysical experiment or neurological examination.

The response of the primary endings to lengthening at a constant velocity includes dynamic and static components. This is illustrated at the top left of the diagram. The continuous line represents the stimulus, a 'ramp stretch' of the muscle. The dotted line traces the curve of 'instantaneous impulse frequency' (ordinates) against time (abscissae) in the Ia axon which carries the signal to the central nervous system. The curve shows that there is a burst of impulses at the very onset of the lengthening, that the frequency of the impulses climbs to a second peak at the end of the lengthening and then falls towards a steady level that measures the new length of the muscle. Thus the primary ending gives a mixed signal of velocity and length-at-any-instant. This ambiguous signal would have to be decoded within the central nervous system by reference to information from other sources.

The response of the secondary endings (Fig. 7, II) to lengthening the muscle shows little dynamic response, and follows the ramp stretch fairly faithfully (Fig. 7, bottom left). In other words it signals only length-at-any-instant.

The mechanical attachment of the primary and secondary endings is to specialized muscle fibres within the spindles. These intrafusal muscle fibres are driven by the fusimotor neurones (γ motoneurones: Fig. 7d, s). Their active contraction stretches, and thereby excites, the primary and secondary endings, but exerts no mechanical action on the muscle as a whole. The responsiveness of the endings to *dynamic lengthening* (velocity and length-at-any-instant) and *static length* of the muscle thus depends on the prevailing levels of activity of the dynamic and static fusimotor neurones. These levels can be adjusted by the central nervous system.

The Golgi tendon-organs (Fig. 7, Ib) are found not only in the tendons but also in the origins of the bundles of skeletomotor muscle-fibres. They are thus coupled 'in series' with the muscle, so that they are excited when it is lengthened passively. Their most vigorous responses, however, are reserved for the active contractions of the skeletomotor bundles which lie 'in series' with them.

It will be clear from Fig. 7 that when a muscle contracts, the spindles, which are coupled 'in parallel' with its skeletomotor fibre-bundles, must be shortened. This would silence the primary and secondary endings, unless the fusimotor neurones were switched on to keep them signalling during the contraction. It turns out that the spindles of contracting muscles—prime movers, synergists, fixators—are *not* silenced, and that their signalling, sustained by fusimotor action, plays an essential part in the automatic governing of movement.

The evidence for the adequate stimuli and response properties of the different types of intramuscular receptors has been gathered over the last sixty years, in particular from accurate quantitative experiments on the ankle-extensor muscles of decerebrate or permanently anaesthetized cats. A single muscle was isolated except for its bony origin and its nerve-and blood-supply; its tendon was attached to a stretcher-myograph for applying ramp and sinusoidal stretches and recording active contractions; the responses of single intramuscular receptors were recorded from single afferent axons dissected from

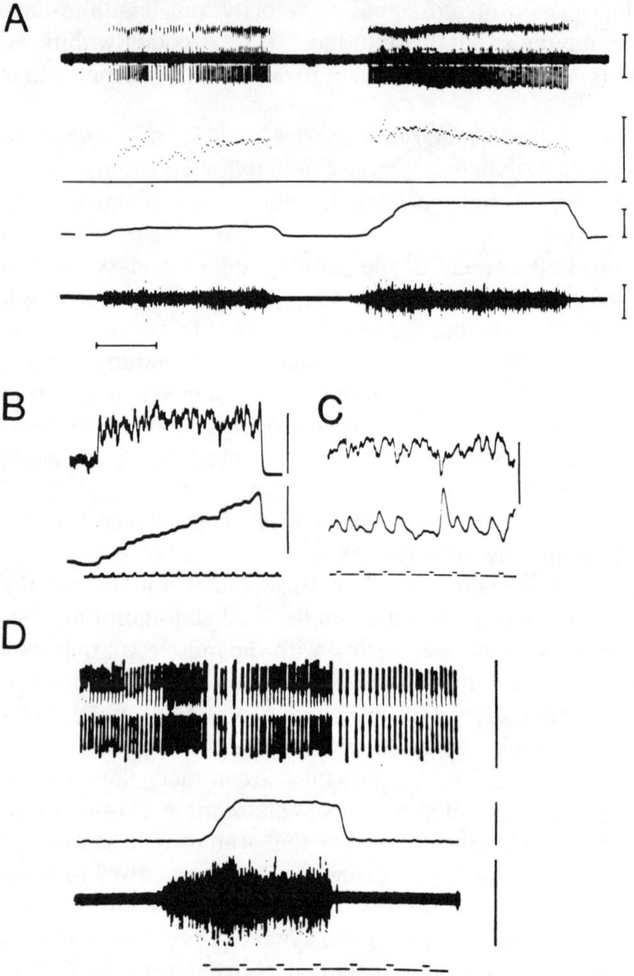

Fig. 8 Microneurography of afferent impulses from muscle spindle of a flexor prime mover of index finger (A) and ring finger (B, C, D)

A, Isometric flexor torque at two strengths. *Calibrations* (from above down): *Spindle afferent impulses*, 60 μV; *instantaneous spindle frequency*, 0–30 impulses per second; *flexor torque*, 0–0.2 Nm; *electromyogram of flexor muscles*, 0.2 mV. Time marker 2 seconds. See text. (Vallbo, *Acta Physiol. Scand.* **78,** 315–333, 1970).

B, Isotonic ramp flexion of metacarpophalangeal joint of ring finger. *Calibrations* (from above down): *instantaneous frequency of impulses from primary ending of flexor spindle*, 0–25 impulses per second; *joint displacement*, 155° to 145° (flexion upwards). Time marker in seconds. See text.

the muscle nerve or dorsal roots, and categorized by their responses to stretch, active contraction and fusimotor stimulation variously combined.

In experiments on one another, begun nearly twenty years ago, workers in the University of Uppsala[7] discovered that they could record from single human axons by gently inserting a fine sharpened tungsten wire, insulated to within 10μm of its tip, into a nerve through the unbroken skin. This is known as microneurography. From the characteristics of the responses to passive movements of joints (passive stretch) and electrically-stimulated contraction of muscles, the identity of the intramuscular receptor responsible for the impulses in a particular axon can be reliably inferred, in the light of the rigorous proofs that had previously been obtained, and could only have been obtained, from the experiments on the exposed muscles and nerves of anaesthetized or decerebrate animals.

Experimenting and thinking about the nervous control of movement owes much to the concept of '$\alpha\gamma$ linkage'.[8] This supposes that in the central nervous control of all but the briefest (ballistic) movements there is a flexibly-coordinated co-excitation of skeletomotor (α) and fusimotor (γ) neurones. Suppose that a ramp movement is proceeding along its pre-programmed course (as in Fig. 8B). The contraction of the prime mover would shorten, or in other words unload, the parallel-coupled spindles within it and would tend to silence their discharge. Suppose that this tendency is being countered by fusimotor driving of the spindles, and that the firing frequency of the spindle receptors is set by the balance between the opposing effects. Then if the ramp movement is slowed by some force that was unpredicted by the programme, the equilibrium will be disturbed. The unloading effect will be checked, the fusimotor effect unaltered, so the spindle

---

Fig. 8 (*contd*)

C, Relation of impulse frequency to small irregularities in the ramp flexion already shown in B. *Upper trace:* faster trace of impulse frequency in second half of record B (time marker in seconds). *Lower trace:* time derivative of joint displacement, i.e. velocity of the ramp flexion. See text.

D, Response of primary spindle ending to isotonic flexion of metacarpophalangeal joint of ring finger, opposed by a steady torque of 0.1 Nm. *Calibrations* (above downwards): 100 μV for *spindle impulses;* 150°–140° for *joint displacement;* 0.2 mV for *electromyogram of flexor muscles. Time marker* in seconds. See text. (B, C and D from Vallbo, in *New Developments in Electromyography and Clinical Neurophysiology*, Vol. 3 (ed. J. E. Desmedt) 1973, Basel: Karger).

signalling will be stepped up. This stepping-up could not only signal that the movement had gone off course, but could also be instrumental in bringing it back (see below, p. 39).

Evidence that $\alpha\gamma$ linkage is actually at work in voluntary movements of the fingers comes from microneurography.[9] Let us look at the results of an experiment. The top two lines of Fig. 8A are the record of impulses discharged by a spindle primary during two voluntary isometric flexions of the index finger. The top line displays the impulses, the second displays the rise and fall of their frequency. The two flexions are recorded in the third line (flexor torque) and fourth line (electromyogram). The second flexion was more powerful than the first.

The spindle discharge began at about the same time as the skeletomotoneurones began to fire. This is proof that there was coactivation of fusimotor neurones: skeletomotor excitation by itself would have shortened (unloaded) the spindle and this primary ending would have remained silent.

In this isometric experiment the finger was held stationary and the muscular shortening was limited to that allowed by the elasticity of the tendon and the fascial part of the muscle's origin. Fig. 8D shows a record of *isotonic* flexion of the ring finger at the metacarpophalangeal joint (10° in 1 s). In the initial position the finger was maintaining a small isometric flexor torque, and the background discharge of this primary ending (top line, Fig. 8D) would have been due to the tonic $\alpha\gamma$-linked fusimotor activity. The isotonic flexion did not reduce the spindle firing significantly, proving that the large unloading effect had been offset by increased fusimotor action.

Alpha-gamma-linked fusimotor action keeps the spindles signalling throughout the movement, but what parameters are they signalling? Fig. 8B and C show an experiment in which the spindle ending was evidently signalling the *velocity* of a ramp movement.

The lower line of Fig. 8B traces a slow voluntary isotonic flexion of the metacarpophalangeal joint of the ring finger which covered an angular distance of 10° in 15 s. The linear ramp was disturbed by slight mechanical irregularities. The upper record is of the impulse-frequency of a primary ending, recorded microneurographically from the median nerve. This ending started abruptly to discharge at about 25 Hz and then oscillated about the same value throughout the movement. Evidently it was signalling the mean velocity and the small irregularities that were superimposed on it (Fig. 8C).[10] The spindle primaries are highly sensitive to very small mechanical oscillations.

So possibly the balance of static and dynamic fusimotor control had been so adjusted by the programme of this elemental movement as to enable this primary ending to 'look at' the velocity. The signal of velocity could have been fed back to the central nervous control loops, and also to the programme itself (Fig. 2), so that any mismatch between the prescribed (Sollwert) and accomplished velocity (Istwert) could be corrected.[11]

Perhaps the signalling by a single spindle primary is not of great quantitative significance. But a significant fraction of the spindles of a prime mover could be signalling in the velocity-sensitive mode. Another population of primaries and secondaries, adjusted possibly by a pre-programmed preponderance of static over dynamic fusimotor activity, could be 'looking at' changes in length, in other words, in joint angle. These could signal any mismatch between prescribed and accomplished angle at successive stages in the 'ecphoration' of a movement.

4. *Segmental and suprasegmental control of a voluntary ramp movement*

From the effects of deafferentation, it can be assumed that afferent impulses generated by the spindles of muscles contracting in α-γ-linkage must normally play an important part in correcting small drifts of a ramp movement away from its programmed trajectory, and in bringing it back on course when it has been deflected by an external perturbation. It might be assumed, further, that these afferent impulses contribute also to our perception of the movement. The first assumption will be considered here and in Chapter IX; the second in Chapter VII.

The simplest of the control circuits of the spinal segments is outlined in the upper half of Fig. 7. The large filled circles represent the pools of skeletomotoneurones of the agonist and antagonist muscles. The monosynaptic and disynaptic connections have been worked out by intracellular recording of excitatory and inhibitory post-synaptic potentials evoked in the skeletomotoneurones by selective stimulation of Ia afferent axons, mainly in the lumbosacral segments of the cat. Such work as has been done in other segments and in other mammals has established the generality of the segmental circuits outlined in Fig. 7. The densest monosynaptic projections from the primary spindle afferents of a particular muscle go to its own ('homonymous') skeletomotoneurones (Fig. 7). In clinical work the integrity of this circuit is routinely tested with a knee-jerk hammer. But there are also 'heteronymous' connections to the motoneurones of other muscles whose postural and locomotor functions are synergic with those of the homonymous muscle. Between cats and baboons there are important differences in the pattern of Ia connections in the hind limb which are related to their different modes of locomotion, digitigrade in cat, plantigrade in baboon.[12]

The Ia axons branch to supply monosynaptic excitation to the Ia-inhibitory interneurones (Fig. 7, top right) which, in turn, monosynaptically inhibit the skeletomotoneurones of the antagonist muscle. This is Sherrington's 'reciprocal innervation' at its most elemental level.

So when a muscle, contracting under α-γ-linked control, is 'unexpectedly' slowed, and its spindles signal this slowing by stepping up their firing in the manner already described, this signal will increase their auxiliary monosynaptic driving of the 'homonymous' skeletomotoneurones (and also some 'heteronymous' ones), and will also increase their driving of the Ia-inhibitory interneurones which inhibit the skeletomotoneurones of the antagonist. This has been felicitously called 'servo-assistance of movement.'[13] Conversely, an 'unexpected' acceleration of the movement will automatically reduce the auxiliary driving of the prime mover and apply less disynaptic inhibition to the antagonist.

Fig. 7 neglects the central destination of the signals from the spindle secondaries (II). There is evidence that they, too, make some monosynaptic connections with skeletomotoneurones. Fig. 7 also neglects the polysynaptic spindle circuits which would be important in sustaining the classical tonic stretch reflex on which much of the servo-assistance of movement must surely depend.

Finally, the input from the Golgi tendon organs disynaptically inhibits the skeletomotoneurones of the prime mover and so can control its output of force.

The skeletomotoneurones of the baboon's finger muscles have been shown by intracellular recording to receive homonymous and heteronymous Ia-excitation,[14] and it can be assumed that this is also the case in man. A finger-jerk might therefore be expected, and is, indeed, a common finding in patients with cortical lesions, in which the skeletomotoneurones are presumed to be facilitated by an excess of background excitation over background inhibition due to the pathologically-unbalanced activity of cerebrospinal tracts. But it is unusual to find a finger-jerk in a healthy person, though the triceps, biceps and brachioradialis jerks are usually as readily elicitable as the knee-jerk.[15] One can imagine that at rest, the motoneurones of these heavier proximal muscles would be kept near to their firing threshold, and thus responsive to the brisk stretch of a tendon tap, by a net excess of background excitation over background inhibition from sources other than their own muscle receptors; sources related especially to posture. Still fingers would lack that sort of background, but moving fingers would have other sources of excitation, especially cutaneous and corticospinal, to interact with the monosynaptic Ia input in the control of their performances.

The responsiveness of the skeletomotoneurones of flexor pollicis longus to stretching that muscle can be tested with the apparatus described in connection with Fig. 4. FPL holds the distal phalanx of the thumb stationary against a steady isometric extensor force imposed by the torque motor. By suddenly extending the interphalangeal joint the motor rapidly stretches the muscle. This excites its intramuscular receptors and injects a nearly-synchronous volley of impulses into the spinal cord. The reflex motor volleys discharged by the motoneurones in response to this unnatural artificial input are recorded electromyographically.[16]

With the most rapid stretches a majority of people show an electromyographic response at a latency of about 25 ms. This corresponds to the time taken by a spinal monosynaptic reflex to travel round the segmental circuit from the spindles to the skeletomotoneurones and thence back to the muscle (Fig. 7). This is the feeble homologue in FPL of the robust 'tendon jerk' in biceps. Under similar experimental conditions biceps always gives a strong response at segmental latency.

To a large, slower, sustained stretch, FPL responded in every subject with a burst of activity beginning 40 ms after the start of the stretch. This is the first component of the so-called 'long loop reflex' (A, Fig. 9). Evidence that its circuit includes the cerebral motor cortex will be presented in Chapter IX. It is purely automatic, entirely outside the subject's voluntary control. Whether he is asked to prepare himself (a) to maintain the stationary position, (b) to pull against the imposed force when he feels it, or (c) to relax when he feels the pull, he cannot influence this first 'long loop' response. The response is not powerful enough to compensate for the perturbation, but would presumably be effective in correcting small deviations from a programmed trajectory. Experiments in which the skin and joints of the thumb are anaesthetized prove that this response originates in the intramuscular receptors.

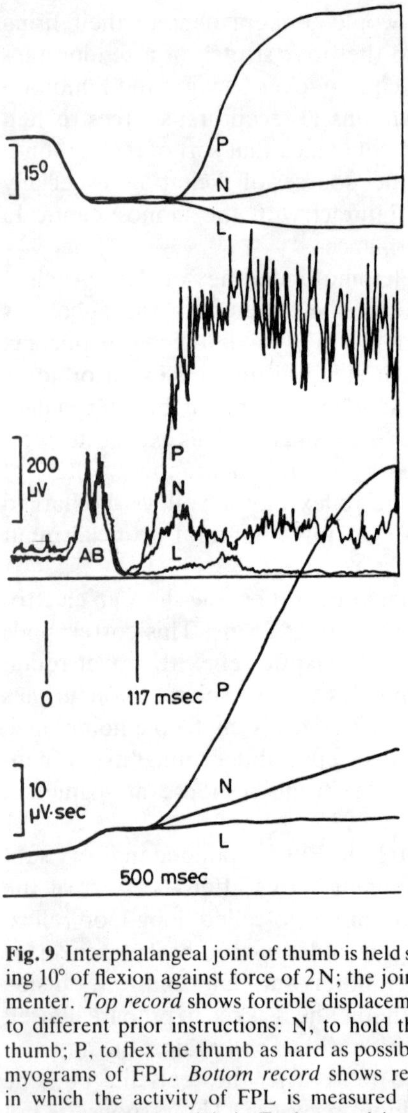

Fig. 9 Interphalangeal joint of thumb is held stationary by the subject, who is maintaining 10° of flexion against force of 2 N; the joint is then forcibly extended by the experimenter. *Top record* shows forcible displacement, followed by responses of the subject to different prior instructions: N, to hold thumb in steady position; L, to relax the thumb; P, to flex the thumb as hard as possible. *Middle record* shows rectified electromyograms of FPL. *Bottom record* shows rectified and integrated electromyograms, in which the activity of FPL is measured by the areas under the curves (steeper slope = increased activity). Each record is average of 48 trials. Time scale 500 ms. See text. (Marsden *et al.*, in *Cerebral Motor Control in Man: Long Loop Mechanisms, Progr. Clin. Neuro-physiol.* **4,** ed. J. E. Desmedt, 167–77, 1978, Basel: Karger)

The second component of the 'long loop reflex' (B, Fig. 9) starts 55 ms after the start of the stretch. This component is slightly, but significantly, increased if the subject has been instructed to resist when he feels the pull, and decreased if he has been instructed to relax. It includes a response at 60 to 70 ms latency which is due to stimulation of the skin and possibly of the joint.[17]

The latency of the final response (117 ms in Fig. 9) varies from one subject to another. In a series of subjects the average latency was 100 ms, the shortest was 85 ms—considerably shorter than any familiar voluntary reaction time (e.g. 150 ms for a voluntary response to an auditory stimulus). This response is clearly determined by the prior instructions given to the subject: to maintain the position of the thumb (N, Fig. 9), to flex it as hard as possible (P), or to relax (L). But how far this is to be attributed to an immediate *voluntary* response with such a short reaction-time, or to a 'gate' in the brain which is 'set' by prior instruction so that the response, when it comes, remains mainly automatic, is a question that continues to invite research.[18] Is the cerebellum the gate keeper? (Chapter IX).

Full compensation is not completed in less than 0.5 s from the perturbation, and then by conscious voluntary correction. In those trials in which the perturbation was most nearly compensated by responses A and B the final voluntary correction was smaller, and *vice versa*.[19]

The foregoing experiment may be compared with the work of an engineer, probing the structural integrity of mechanoelectrical networks. But how do these networks actually function during a ramp movement? If, during a ramp flexion of the interphalangeal joint of the thumb, the movement is suddenly halted by the torque motor, the increased electromyographic activity in FPL starts during the period of the engineer's long-loop response (AB in Fig. 9). There is no response at segmental monosynaptic latency. Fig. 10 shows the responses in two subjects. In subject B the increased electromyographic activity (*Halt*, upper trace) began 60 ms after the perturbation (*Halt*, lower trace). The latency in subject A was 50 ms. Since the muscle was never stretched, but was merely halted, this is clearly a case of servo-assistance in action.[20] The reduction of e.m.g. activity in response to sudden removal of the steady torque that was resisting the ramp movement (Fig. 10, *Release*, and the increased activity in response to sudden stretch of FPL by forcibly extending the voluntarily-flexing joint (Fig. 10, *Stretch*), are also servo-type responses, occurring at the same latency.

These beautiful records of servo responses to perturbation of volun-

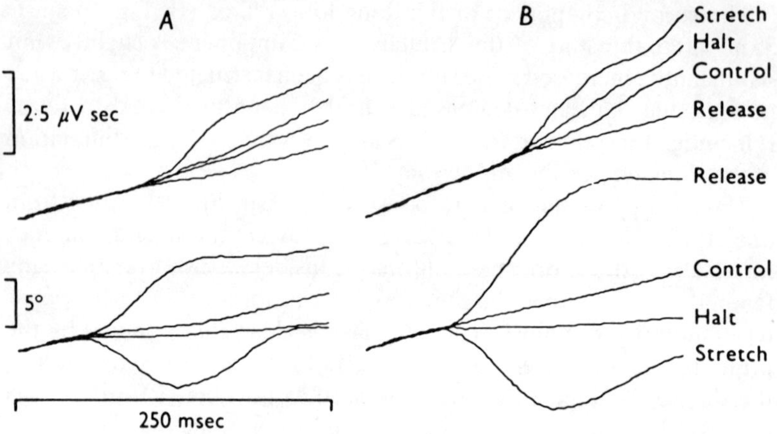

Fig. 10 Servo-responses of two subjects, A and B, to perturbations of ramp flexion of interphalangeal joint of thumb. *Upper records:* rectified and integrated electromyograms; *lower records:* displacements imposed on the flexing joint. Each trace is average of 16 trials. See text. (Marsden *et al.*, *J. Physiol. Lond.* **257**, 1–44, 1976)

tary ramp movements are known familiarly as 'tulips'. The most labile part of a tulip is the *Halt* response, whose latency sometimes exceeds that of the more robust responses to *Stretch* and *Release*.

CHAPTER V

'The muscular sense'

We pass now from the performance to the perception of movements. Chapter IX of Sir Charles Bell's Bridgewater Treatise on the Hand is headed 'Of the Muscular Sense'. 'We are sensible of the position of our limbs,' he wrote, 'we know that the arms hang by the sides; or that they are raised and held out; although we touch nothing and see nothing. It must be a property internal to the frame by which we know this position of the members of our body: and what can this be but a consciousness of the degree of action and the adjustment of the muscles? At one time I entertained a doubt whether this proceeded from a knowledge of the condition of the muscles or from a consciousness of the degree of effort which was directed to them in volition'. (He says nothing about possible knowledge of the condition of the skin and joints). He then says that it was this doubt which led him to discover the sensory and motor functions of the dorsal and ventral spinal nerve-roots.[1] 'I had reasoned in this manner—we awake with a knowledge of the position of our limbs: this cannot be from a recollection of the action which placed them where they are; it must, therefore, be a consciousness of their present condition. When a person in these circumstances moves, he has a determined object; and he must be conscious of a previous condition before he can desire a change or direct a movement.' Bell called this consciousness 'a sixth sense'. The term 'Kinaesthesis'[2] has often been applied loosely to all aspects of this consciousness, static as well as kinetic.

The patient suffering from distal sensory neuropathy re-enters at this point. He was quite unconscious of the positions and willed movements of his hands when his eyes were closed. His intact motor apparatus failed to give him any 'consciousness of the degree of effort which was directed to them in volition' to tell him what was going on. He could not judge weights. When carrying a suitcase he had to keep looking down to make sure he had not dropped it. Even with vision he could not pick up small coins, particularly off a soft surface, or do up buttons. He could not write, because he could not control the pressure of the writing point on the paper. To hold a cup he had to use both hands. His hands got into dystonic postures when holding a glass, knife and fork. He remained able to drive his old automobile, even at night, but was unable to learn to drive his new one—unable, that is, to create the necessary new programmes for gear lever, pedals etc. in the absence of any flow of information from his deafferented hands and feet.

It is very striking that the complex programmes of such a skilled performance as driving, once the performance has been firmly transferred from a 'least automatic' to a 'high automatic' level, can be run off without any response feedback from the feet and hands when operating the pedals, wheel and gear lever (Bernstein's 'working organ, instrument, object' Fig. 2). If the surviving feedback from more proximal sources—knee and hip, elbow and shoulder—was necessary for running the old programmes, it was not sufficient for the establishment of new ones.

A specially unpleasant symptom was that when the patient was dropping off to sleep his arms seemed to separate from his body, unless he kept moving them. This recalls the similar experiences of student volunteers in Hebb's experiments on perceptual isolation.[3] Bell would have diagnosed loss of 'knowledge of the condition of the muscles' and cited the case as a fresh proof of the reality of the Muscular Sense.

1. *Definition of 'Muscular Sense'*

Sherrington's chapter on the Muscular Sense[4] defined it as 'all reactions on sense arising in motor organs and their accessories'. From the context it is clear that he meant peripheral motor organs, not motor organs within the CNS. He grouped the perceptions into '(i) those of posture, (ii) of passive movement, (iii) of active movement, and (iv) those of resistance to movement. Changes in consciousness accompany all movements of the body which occur at speeds above a certain liminal, and over distances of beyond a certain liminal extent.' But he went further. 'The muscular sense embraces the part played by sources of afference in the taxis of muscular activity, even where the reactions lie not amenable to introspection.' So he would have included the automatic governing of movement by somatic afferent response feedback which we were considering in the previous chapter. 'Certain elementary adjustments of muscular action occur reflexly without awaking or in the slightest degree seeming to trouble at all the stream of consciousness. Yet these have, inferiorly and remotely, relationship to the highest achievements of the muscular sense, such as the manipulative finesse of the human thumb and finger.'

It seems a plain contradiction in terms to speak of the muscular *sense* in connexion with automatic activities which a conscious person is not aware of, and which he cannot attend to or influence even if he wishes to do so. It is true, as Sherrington said, that 'the physiologist derives a certain advantage from this, in that it offers an objective mode of studying the activities of the afferent apparatus of muscular sense by watching the objective effects of it on the characters of muscular movements.' He was thinking of the paralysis and ataxia that follows the cutting of afferent nerves, and the action of weak electrical stimulation of afferent axons from one muscle, exciting that muscle and inhibiting its antagonist—producing coordinated movements even in the decerebrate preparation. 'Kinaesthesis, in the psychological sense, seems precluded where the whole cerebrum had been removed.' The fact that the same receptors supply the information that is necessary for the automatic governing of movement as well as that necessary for the perception of his own movements by a conscious perceiver does not seem to justify the indiscriminate application of the term 'muscle sense' to these two distinct functions, the one conscious, the other not 'amenable to introspection.'

Only in the most general way is it useful to think of ourselves as possessing 'five' or 'six senses'. People who do research on vision are not really working on a single 'sense', but are actually coming to grips with several distinct problems—the structure of the topographical map of visual space in the brain, the extraction and non-topographical representation of such features as the luminance and colour of objects, their size, and their direction and velocity of motion.[5] All these, however, originate in a single receptor sheet, the layer of rods and cones in the retina. It seems question-begging to speak of 'muscle sense', implying that this originates solely in muscle, as 'visual sense' originates solely in the retina.

It might be well to begin by dismissing the common statement that muscles are 'sentient.' We do feel something when deep pressure is made on them, and in cramp, and after 'pulling' them in sport, we feel severe pain' but these are not the sensations we mean when we speak of 'muscle sense' or 'kinaesthesis.' We mean awareness of the position of parts of our bodies, and of the changes in position that are brought about by passive and active movements of joints; of efforts exerted; of forces achieved; of speed and rate of change of speed; and of the trajectories of movements. All these we refer to internal brain-models of our bodies and of visual space, built up by learning during infancy and childhood.

2. *Receptor organs for perception of positions and movements*

Having distinguished these more-or-less separable perceptions, our practical objective is to try to relate them to signals arising in specific classes of receptors: in the primary and secondary nerve-endings of the muscle spindles, the Golgi tendon organs, the nerve-endings in joints and skin. We must also try to define the contribution of the axon collaterals of corticospinal neurones to these perceptions, especially to the classical 'sensations of innervation', 'sense of effort' and 'felt will'. The concept of 'muscle sense' has done valuable work since Bell's invention of it, but has now, perhaps, earned a rest. 'Perception of movement' (kinaesthesis), 'perception of position' (stataesthesis?), which beg no questions about their sources of information, may be better guides to thinking and experimenting for some time to come.

Analytical isolation of separable inputs in order to define the quality and intensity of the perceptions they can arouse—for example, by stretching muscles, or by vibrating their tendons to excite the primary endings of the muscle spindles—is an essential stage in the investigation, but not the end of it. In real-life situations the brain is likely to 'select' varying proportions of information from the different sources—information arriving along several parallel channels—according to the requirements of some perceptual 'set' or motor strategy. As Granit has put it, 'whatever clue the brain be given, it also is capable of interpreting if need for this arises to serve some sensible purpose. Meaningless pull on muscles is not regarded as serving any such purpose.'[6] So we need not expect to find that isolated excitation of a single source of input will always evoke a recognizable sensation, or give rise to an accurate perception of a position or movement. Indeed, there is much to be learned from the illusory perceptions that are aroused when the separable inputs are dissociated experimentally, and from 'constant errors' in motor performance. Accurate perceptions are likely to depend on complex central interactions between several parallel sources of input and the 'corollary discharges' of corticospinal axon collaterals that can excite and inhibit the first-, second- and higher-order neurones along the somaesthetic pathways leading to the cerebral cortex. Details of such interactions—'gating'—may vary from time to time in different individuals, and in different cognitive and perceptual-motor 'sets', in ways yet to be discovered.

3. *Perception of position and movement at joints of the finger and thumb*

The powerful new method of microneurography makes it practicable, for the first time, to investigate directly the relation between the activity of specific classes of receptors and the perceptions they can arouse in conscious man.

We are perfectly aware of the positions of our fingers at rest. As Sherrington wrote, 'It is apparent to most of us that if, with closed eyes, we attend to the consciousness of the resting arm, the hand, and especially fingers, are the parts we are most aware of, best outlined, so to say, in consciousness.'[7] For the source of this awareness the muscles' length-receptors would seem the obvious candidates. But microneurography shows that there is very little signalling from these at rest, often no signalling at all (as in Fig. 8A).

One might still expect that the spindles could be sensitized by brain-controlled fusimotor activity in a subject who was deliberately 'attending to the consciousness of the resting arm'. Theoretically, very small movements to either side of the prevailing mean position of the joint would then be readily detected by the primary and secondary endings of an antagonistic muscle-pair, and could be 'interpreted' as positional information of the joint angle. But we have no evidence that such fusimotor sensitization actually occurs. Microneurography gives 'no indication that resting subjects at will (by mere intention to move, by imagining imposed movements, etc.) can alter the afferent firing from a relaxed muscle held at constant length or modify the afferent stretch discharges to passive length changes. . . . The fusimotor systems in man seem to be at the service of the spindles in contracting, as opposed to relaxed muscles.'[8]

We turn next to the joints. Their capsules contain nerve-endings, and since our perceptions of positions and movements are referred to the joints we should expect *a priori* that these endings would make a major contribution. That expectation has been disappointed in the case of the hand. Microneurography has isolated only 2 out of 111 axons sampled in the median nerve whose discharges can be attributed to interphalangeal joints, and neither of these fired except at one extreme position of the joint.[9]

The skin is the only remaining candidate, and the contrast is striking: 77% of receptors in the glabrous skin of the hand discharged during movements of the fingers, although their receptive fields were not in contact with any external object, and there was no sensation of contact, only of movement. The skin-creases are evidently 'exoskeletal' joints in their own right, supplying more information than the interphalangeal joints of the bony endoskeleton. But since the same receptors also subserve tactile sensibility, there is (as in the case of the spindle primaries with their mixed signal of position and velocity) an inherent ambiguity and a need for decoding by the central nervous system, by reference to information from other sources.

There are slowly-adapting skin receptors (SAII) whose impulse-frequency is linearly related to the angle of the adjacent distal interphalangeal joint.[10] These could maintain awareness of a stationary position for many seconds on end; thereafter the awareness would be expected to fade, until refreshed by the next small movement. In a paralyzed patient no small movements would be forthcoming. Here Brodal's personal experience of the first weeks of a pure motor hemiplegia is relevant. 'When the fingers had not been moved for some time, the patient, as is usual, had no feelings of his muscles in the hand and forearm and of the position of the finger joints'[11]—in spite of the fact that their sensation, as tested by passive movements, was intact. In health our fingers, in contrast to those of a hemiplegic, are probably never still for more than a few seconds at a time.

4. *Perception of position and movement at proximal and axial joints*

If the relatively weightless fingers depend for position sense on impulses from their skin, it is quite otherwise with the heavy proximal parts of the arm, and this calls for comment. Bell's statement, already quoted, is clearly concerned with knowledge of the postures of the limbs and trunk, and clinical neurologists since Gowers have recognized that these have nothing to do with the skin. 'The sense of posture may be lost when cutaneous sensibility is normal, and perfect when this is much impaired. . . . The chief source of these perceptions must therefore be the deeper afferent nerves, those of the muscles and joints, perhaps chiefly of the muscles.'[12] But keeping in mind the different functions of maintaining postures and carrying out tactile explorations, we ought not to be surprised if the relative contributions of muscle and skin to our knowledge of resting posture were to differ as between shoulder and fingers. In terms of threshold excursion and threshold velocity, our perceptions of passive movements at the shoulder are about forty times more delicate than those of the fingers. The lower threshold for consciousness at shoulder mirrors its low reflex threshold and powerful response to displacement in the gravitational field on which the steadiness of the arm depends and which, in turn, supplies the firm postural foundation on which the movements of the fingers are based. For perception and posture at proximal and axial joints the muscle spindles are of leading importance.

CHAPTER VI

'The muscular sense': perception of passive movements of the fingers

Even if passive movements of the fingers play little more part in everyday life than, say, the knee jerk, psychophysicists and clinical neurologists are greatly interested in them for the light they can throw on the receptors and afferent pathways that are necessary for the perception of movement.

The joint receptors are evidently unimportant. They are too rarely encountered in microneurography, and they supply no information of joint angle between the extremes of flexion and extension. And patients whose metacarpophalangeal joints have been replaced by flexible silicone-rubber implants show no loss of awareness of passive movement.[1]

Slowly-adapting receptors in the skin of the fingers are frequently encountered by microneurography during *active* movements of the fingers, even when there is no contact with an external object. They give an accurate signal of joint angle. Their importance in the detection of *passive* movement seems self-evident. But the routine clinical testing of passive movement, in which the examiner grips the sides of a distal phalanx and flexes and extends the interphalangeal joint, cannot fail to excite the slowly-adapting skin receptors. The only way in which the complicating effect of skin contact can be excluded is by *centripetal* pulling on an exposed finger tendon—hardly a routine clinical test! It will be seen below that such pulling, at the level of the wrist, arouses a vivid perception of angular movement at the joint. In view of the impoverished information from the joints, the perception in this case probably arises largely from the skin, and possibly also from receptors in the intrinsic muscles of the hand which might be affected by the tendon-pulling at the wrist.

1. *Muscle spindles and passive kinaesthesis*

The lack of contribution of muscle spindles in the *extrinsic* finger muscles in the forearm to position sense ('stataesthesis') in the fingers has already been noted. The possible contribution of their length-

and velocity-measuring receptors to passive *kinaesthesis* must now be considered. Passive lengthening of one of the muscles in Figs. 7 and 11 by moving the joint should excite the primary and secondary endings of its spindles.[2] If any sensation were aroused, it would be expected that the subject would feel that the joint had been moved in the direction that would lengthen the muscle—say, a sensation of extension when the flexor muscle was lengthened by passive extension of the joint. Lengthening of the flexor *ipso facto* shortens the extensor muscle, so that if the extensor spindles had been discharging the shortening would unload them and silence their discharge. The combined effect would contribute to the sensation that the joint had been extended by the examiner.

By 'ring-blocking' a finger by injecting local anaesthetic around its base, or by making the whole hand ischaemic by inflating a pressure-cuff round the wrist, the stimulation can be limited to muscle receptors—those of the extrinsic muscles whose bellies are in the forearm. Kinaesthesis can then be tested by imposing passive movements on a digit or digits deprived of receptors in skin and joints in the case of the ring-blocking experiment, and additionally, in the ischaemic experiment, of receptors in the intrinsic muscles of the hand. In a seminal series of experiments[3] the subjects were required to imitate with the opposite hand the movements they perceived in the experimental hand, as well as describing their introspections.

In the experiments on locally-anaesthetized fingers and ischaemic hands there was always a less-than-normal awareness of passive movement when the muscles were relaxed. Below a speed of 1°/s no sensation was aroused. Speeds of 5° to 10°/s enabled reliable detection. Awareness was improved by voluntarily tensing the long muscles. The quality of the kinaesthetic sensation was normal. At the interphalangeal joint of the thumb there was marked variability between subjects: with the long muscles relaxed, some failed to detect 90° of movement, others had no difficulty. With tensed muscles all subjects could detect the imposed movements. The voluntary contraction would have been accompanied by increased fusimotor action on the spindles, increasing their responsiveness to lengthening and unloading; and also by increased excitation of the Golgi tendon organs of the contracting muscles (see Fig. 7).

Ingenious use has been made of the anatomical arrangement of the long (extrinsic) muscles of the hand, whose tendons work the distal interphalangeal joint of the middle finger. When the index, ring and little fingers are kept fully extended at all joints, and the middle finger is fully flexed at the proximal interphalangeal joint, the muscles acting on the distal joint are 'disengaged'—so slackened that they cannot move the joint; nor can they be lengthened, nor further shortened, when the experimenter passively flexes and extends the joint. Passive movements of 10° were imposed on the joint at angular velocities ranging from 1° to 27°/s, and the scores for detection of movement (including direction of movement) with the long muscles 'disengaged' (inputs from skin and joint only) were compared with the scores with the muscles 'engaged' (inputs from skin-plus-joint-plus-muscles). By anaesthetizing the middle finger, the detections by the long (extrinsic) muscles could be scored separately. These scores varied widely. In some subjects 'muscle sense was very poor, and even displacements made at very fast angular velocities were not reliably detected.' The experimenter[4] comments: 'This variability may explain how some confusion arose in earlier experiments concerning muscle afferents and perception—if subjects with very poor muscle sense were chosen for testing it is not surprising that the sense was said to be absent.' In some subjects 'muscle sense' scored higher than 'joint plus cutaneous sense', and was almost as good as 'the full kinaesthetic sense' (joint-plus-skin-plus-muscle). Again, 'muscle sense' was always improved by voluntarily tensing the long muscles, presumably by sensitizing the spindles by fusimotor activity.

2. *Skin and kinaesthesis*

The probable necessity, for correct perception, of having more than one source of information is underlined by these elegant experiments. 'In many subjects poor muscle sense and quite average joint-plus-cutaneous sense combine to give a complete position (sic) sense that is superior to either sense alone'. Central integration of the three inputs could conceivably be variously adjusted in different perceptual-motor sets, according to the wording of the instructions given to the subjects and their understanding of what they are being invited to achieve.

Clinical experience in hand surgery,[5] long antedating the discoveries of microneurography, points to the importance of cutaneous receptors for the awareness of passive movements of the digits. In cases of nerve injury in which a finger is denervated with sparing of its extrinsic muscles, it is found that passive movements of that finger are not perceived. This is so in patients in which the interossei of the anaesthetized middle finger, as well as its three extrinsic muscles, were preserved. In patients who had had a flap of innervated skin transferred to the volar surface of a denervated finger, perception of passive movements of that finger was restored.

Collateral experiments were also made on normal subjects. These differed from those reviewed on p. 49 in that anaesthesia of the whole hand distal to the wrist was produced by blocking the ulnar and median nerves with local anaesthetic and infiltration of the skin containing the dorsal radial nerve branches, all at wrist level, instead of by ischaemia. It was found that passive movements of the index and little fingers went unperceived, but that there was some residual perception of those of the thumb. Individuals varied considerably in their awareness of passive movements of the thumb. These movements caused visible deformation of the skin overlying the tendons of the anatomical snuff-box proximal to the level of anaesthesia. It was suggested that such deformation has possibly been an unrecognized source of error in experiments designed to isolate the contribution of muscle receptors to the perception of passive movement.

3. *The tendon-pulling experiment*

To find out whether passive lengthening of an extrinsic finger muscle arouses a sensation of movement at the corresponding joint, the simplest conceivable experiment would seem to be to expose its tendon and pull it, taking care not to move any structure within the hand. The subject's ability to detect the pull, to describe the sensation aroused and to imitate the perceived movement by moving the corresponding joint or joints in the opposite hand can then be tested. The experiment can be done, with their informed consent, on patients undergoing carpal tunnel surgery and other operations on the hand.

Two of the leading investigators have been sufficiently interested in getting the answers to supplement their clinical observations by the heroic expedient of having the experiment performed on themselves: D. I. McCloskey on his extensor hallucis longus[6] and E. Moberg on his flexor digitorum superficialis.[7] The fact that two such skilled observers have disagreed about the interpretation of one another's results—though not about the results themselves—shows straight away that the experiment cannot be as simple as was expected *a priori*, and that further thought needs to be given to its design before repeating it on patients and on healthy volunteers. The publication of these fascinating papers side-by-side in *Brain* will be helpful to anyone who wishes to take the matter further, and feels a proper obligation to analyze them carefully before doing so. Although comments by a writer without practical experience may well seem impertinent, I shall venture a few. These may begin with the suggestion that the experiments on perception of passive movements that have been outlined already should prepare one to expect considerable variability between the sensitivity in different subjects; and that any sensations evoked will, even at best, fall short of full normal kinaesthesis. These considerations suggest that the tendon-pulling experiment is not necessarily going to give a simple yes-or-no answer.

In so far as the pulling of an extrinsic flexor tendon can elicit a sensation of extension of one or more joints the two series of observations are in agreement. This result was obtained in three of Moberg's six patients and in all four of those of McCloskey and his colleagues. Disagreement is about the amount of muscle lengthening that is legitimate. Moberg's measurements have convinced him that sensations are aroused only by pulling a muscle beyond its maximum possible length in the body, when the pulling comes up against a 'stop'. He concludes that with wrist and fingers held in a straight line the muscles are already practically at their maximum permissible length; and that the maximum physiological length, with wrist fully dorsiflexed, is only 6 to 8 mm longer. In the range between the minimum and maximum physiological length, that is, with the joints partially flexed, he finds that a tendon can be pulled as much as 40 mm into the wound without eliciting any sensation; and that the sensation elicited by an excessive pull may be accompanied by visible grooving of the skin overlying the muscle belly, and can be abolished by extensive local anaesthesia of the cutaneous sleeve of the forearm. The digital joints of the four patients observed by McCloskey were all held in full extension, and the wrists were held straight. Disputing the validity of Moberg's measurements, McCloskey states that full dorsiflexion of the wrist lengthens the digital flexors a further 10 to 15 mm, and that this correctly represents their maximum physiological length. He finds it difficult to be sure that significant muscle stretch occurred in Moberg's patients. Moberg, for his part, considers that McCloskey *et al.* should have eliminated the possible contribution of receptors in the skin of the forearm by extensive injections of local anaesthetic.

In the experiment on the flexor superficialis tendons to Moberg's own index and middle fingers, centrifugal pulling (roughly sinusoidal, at 0.3 to 1.0 Hz) within the physiological range of length elicited no sensation. Pulls which reached the 'stop', some of which were seen by the surgeon to cause 'a slight movement of skin proximally in the forearm,' all elicited sensations proximally in the forearm. 'The consciousness of finger extension in this situation—there was never any feeling of finger flexion—seemed to the subject as much an interpretation of what was felt as a direct signal from moving fingers. The experimenter of course saw that the digits were lax and immobile. The subject could not distinguish between pulls of the superficialis tendons of the index and the middle fingers.' The effect of anaesthetizing the cutaneous sleeve of the forearm was not, however, tested. In contrast, 'the slightest centripetal pull, flexing a digit only a few degrees, was immediately reported as coming from the correct digit and from its proximal interphalangeal joint. The subject could tell the experimenter that he was working on the superficialis tendon of index even before he had identified it in the wound. Later, when the experimenter thought, correctly, that the subject was expecting a pull on the ring finger, he tested the index finger instead, and was immediately told that it was now the index. The overwhelming difference between the quality of the sensation obtained from centripetal pulling and the occasional vague sensation from pulling in the other direction, where little movement of the skin occurred, must be stressed.' As has been mentioned already, centripetal pulling of an exposed tendon should passively excite the slowly-adapting cutaneous receptors of the finger without the complicating factor of contact with the skin.

The interphalangeal joint of McCloskey's big toe could be moved passively through 70°, which corresponded to 7 mm movement of the tendon of extensor hallucis longus (EHL). After this correspondence had been determined, and length markers positioned, the tendon was cut and its proximal end sewn to a sterile rod which connected it to an electromagnetic puller. In response to lengthening, plantarflexion of the interphalangeal joint was regularly reported. The muscle was exerting a tension of 400 g at the measured resting length of the tendon, and when this tension was suddenly released, there was a sensation of dorsiflexion of the joint, due, one imagines, to the sudden shortening (unloading) of the spindles of EHL. 'Sinusoidal displacements of the tendon at 1 Hz or 5 Hz were felt as to and fro movements of the toe at its terminal joint. All sensations associated with tendon pulling were clear, but were noted to be unusual because they were unaccompanied by sensations of cutaneous distortion and pressure *in* the toe itself. At the time of the experiment the sensations were described as being "as if the toe is blowing back and forth in a breeze."'

The methods of measurement of thresholds for detection of sinusoidal and ramp stretching are a model for any future investigation of tendon-pulling. These threshold movements, 0.35 to 0.8 mm in extent, were equal to those for the equivalent threshold angular displacements of the joint of the intact toe, as measured a few days before the tendon-pulling experiment. McCloskey comments: 'at the conclusion of data collection, when the subject was permitted to sit forward and look directly at the tendon in his foot while such displacements were imposed, he could not *see* the movement of the tendon and its connections which elicited the sensations. Neither the subject nor the experimenters noted any dimpling of the skin over the muscle belly in the leg. It seems almost inconceivable that such small stretches could have produced their very specific sensory effects by exciting remote cutaneous or fascial receptors.' Such remote effects were not however formally excluded by enlarging the area of cutaneous anaesthesia proximally to the 6 cm diameter circle, centred on the metatarsophalangeal joint, established for the surgery; but the $2\frac{1}{4}$ hours of experimentation were evidently fully enough occupied to justify the omission. The experiment was concluded by suture of the tendon and closure of the wound. The toe was restored to normal function in a few weeks.

In the 'centrifugal' tendon-pulling experiment a muscle is lengthened at speeds within the normal everyday range. All investigators agree that the resulting sensations fall short of the normal kinaesthetic experience, in contrast to those evoked by pulling the tendon centripetally. It has been well observed that 'in such peculiar circumstances a subject may easily fail to concentrate sufficiently to become aware of a sensory experience that is admittedly of low "intensity".'[8] To overcome this difficulty it would seem necessary to employ a strict experimental protocol. Thus a light or sound signal would warn the subject that he might or might not feel something during the ensuing five seconds. Stretch and non-stretch would be randomized, and positive, false-positive and negative responses would be scored. After each positive or false-positive *detection*, the *perception* would be imitated by movement of the corresponding finger of the opposite hand.

The disagreement on the interpretation of the tendon-pulling experiment centres on the amount of lengthening that is legitimate, and on the need to exclude unwitting stimulation of the skin—by traction on skin overlying muscle-bellies which arise in part from the deep fascia, by raising the skin overlying pulled tendons, or by frictional drag of sliding tendons acting on its under-surface. These complicating factors can be excluded only by very extensive cutaneous anaesthesia. But what is still lacking, to my knowledge, is any *positive* information of the threshold and quantity of spindle activity evoked by tendon-pulling over different ranges of length and at different speeds. Clearly it is likely to be much less than the massive barrage of Ia impulses that can be evoked by vibration of a tendon (see below, p. 56). The obvious experiment, one combining tendon-pulling with microneurography, may seem altogether too heroic to contemplate.

The sensitivity of McCloskey's extensor hallucis longus to very small lengthenings seems, on the face of it, superior to anything so far reported from tendon-pulling in the hand. A leading theme of the lecture was that our perceptions of the postures and movements of different parts of the body, e.g. fingers and shoulder, deserve to be examined separately rather than forced into a single, generalized mould in which the relative contributions of the several parallel inputs—muscle spindles, tendon organs, joints, skin—are supposed to be everywhere and always unvarying. Generalization is an essential scientific procedure, but there is a danger that important differences in function may get overlooked. Confining our attention to the input from the muscle spindles, quantitative comparisons of the detection-thresholds for pulling on the exposed tendons of extensor hallucis longus and flexor digitorum superficialis—measurements of Sherrington's 'speeds above a certain liminal, and distances beyond a certain liminal extent'—are evidently desirable. But the opportunities for data collection may be few and far between, and because of possible individual variation, consistent results may turn out to be as elusive as a will-o-the-wisp. Yet consistent results from many individuals would be needed to show whether the spindles' access to the 'sensorium' is actually more powerful from the big toe and weaker from the fingers, at least under the 'peculiar circumstances' of the experiment.

A priori it might be thought possible that the different functions of the hands and feet would be reflected in different degrees of responsiveness to the isolated spindle input. The toes have important postural and locomotor functions which the fingers hardly share. The mobile thumb and index are the macula of the somaesthetic system. The automatic control of movements of the big toe, and the perceptions that can be aroused from it, could depend proportionately more on the spindles than on the skin. The converse could be true of the fingers. But it could be quite otherwise in the toes of people born without arms and who have grown up to use a knife and fork, to write, draw, and knit with their feet.

4. *Vibrating the primary spindle receptors arouses kinaesthetic illusions*

Up to this point, discussion has been grappling with two questions: first, whether signals from the muscle spindles are *capable* of arousing any perception of movement at a finger joint; second, the extent to which the spindles *actually contribute* to our perceptions of passive movements. But if there is still uncertainty about the answer to the second question, there is none about the first. Animal experiments have proved that longitudinal vibration of the tendon of a relaxed muscle is a potent and highly selective stimulus to the dynamically-sensitive primary endings of the muscle spindles.[9] Percutaneous vibration (e.g. 0.5 mm at 100 Hz), applied to a human tendon, transmits a longitudinal component of vibration to the muscle-spindles. In the absence of vision this arouses vivid illusions of movement.[10] Application of the vibrator to adjacent skin, bone or joint arouses no illusions. The preponderant action is evidently selective, on the dynamically-sensitive primary endings, but microneurography has shown that there is some additional excitation of secondary endings, tendon organs, Pacinian corpuscles and receptors in the skin. 'The minuscule stretch of vibration which has no counterpart in nature',[11] when applied to a flexor tendon, elicits an illusion of extension of the appropriate joint, and *vice versa*.

Percutaneous transverse vibration, as a tool, is more readily applicable to large tendons than to small. Much has been learned from it about the perception of illusory movements at the elbow joint when the biceps and triceps tendons are vibrated. If the forearm is suspended in a horizontal plane so that gravity does not act on the joint, vibration of the biceps tendon, so adjusted in amplitude and frequency as not to cause any reflex contraction of the muscle, creates the vivid illusion that the elbow is extending (as if the biceps were being lengthened). The direction and velocity of the illusory movement can be determined by getting the blindfolded subject to match it with a voluntary movement of the opposite arm. The illusion can continue as long as the vibration is kept up, for a minute or more; it does not correspond to any possible real movement at the elbow-joint at anything like the illusory velocity. Vibration can also create the illusion that a joint has been moved into an impossible position: vibration of the wrist flexor bellies at the elbow seemed to the subject to move his hand into an anatomically-impossible state of hyperdorsiflexion. Such can be the effects of a massive isolated input from the primary endings of the muscle spindles, uncorrected centrally by other inputs coming in along parallel channels or by impulses descending from the brain.

In contrast to biceps, vibration of the tendons of extrinsic muscles of the hand may fail to arouse any illusions of movement, presumably because the simultaneous vibration of several muscles, intrinsic as well as extrinsic, sends additional and conflicting signals to the brain. But when the whole hand was numbed by ischaemic anoxia, vibration of flexor pollicis longus aroused the illusion that the thumb was extending; vibration of EPL and EPB aroused the illusion that it was flexing. These experiments leave no doubt that impulses generated by the spindles of 'isolated' finger muscles arouse perceptions, and that, when these are the sole or major input, the perception is of angular movement of a joint in the direction that would passively lengthen the muscle.

Attempts at longitudinal vibration of exposed human tendons have so far been disappointing. Such an attempt, by researchers whose tendon-pulling on thumb and finger flexors in five surgical patients aroused sensations which were always referred to the correct digits, got no results; but the researchers were uncertain about how effectively the vibration was being transmitted along the tendon to the muscle.[12] The experiment on McCloskey's big toe was bedevilled by 'an intense buzzing sensation, like a tight sock, pressing over the whole foot.' When this was minimized by limiting the vibration to 20 to 100 μm at 50 to 100 Hz—although it was 'still intense'—there was often a perception of plantar-flexion at the interphalangeal joint, at velocities (imitated by the subject by moving a hand-operated lever) of 30° to 70°/s.

CHAPTER VII

'The muscular sense': perception of active movements

We come now to the perception of voluntary performances of the hand. For the perception of movement (kinaesthesis), all the receptors that are excited by passive movements—skin, joints, spindles of muscles passively lengthened or shortened by the movement—would be available. Two further perceptions have still to be considered.

The first is the perception of *force achieved* ('dynamaesthesis' if you like). The Golgi tendon organs of the prime-mover muscles, with their selective sensitivity to active contraction, would supply specific information of force.

The second is the perception of *effort exerted*. One obvious difference between voluntary and passively-imposed movements of the hand is the activity of those corticospinal neurones which drive the segmental motor apparatus of the brachial enlargement of the spinal cord. Collateral branches of their axons have been shown to excite and inhibit 'sensory' neurones[1] which receive and process the inputs from the peripheral receptors and project them towards the 'sensorium'. Some mode of interaction between the motor and somaesthetic pathways, not yet localized or understood, could in theory arouse a perception of effort that was proportionate to the intensity of the motor discharge—a perception corresponding to the classical 'sensation of innervation' or 'felt will'.

1. *A conflict between kinaesthesis and motor control?*

But before coming to the perception of *force* and *effort* there is a question about the perception of voluntary *movement*. It is about the possible kinaesthetic effects of the spindles of muscles actively engaged in the movement—as prime movers, synergists and fixators.

When a prime mover contracts, its spindles are shortened, and if nothing else happened their firing of impulses would decline or stop. This could be interpreted centrally as a signal of movement of the joint in the direction-of-action of the prime mover. It would be consistent with the signalling from the passively-lengthened spindles of the antagonists, from the skin, and with any signalling from the joint itself (Chapter VI).

But the spindles of the contracting muscles do *not* fall silent. They are kept firing by the fusimotor neurones in the service of the automatic control of movement by α-γ-linkage (Chapter IV). We know that their primary endings, when stimulated near-maximally and in virtual isolation by vibration, arouse a vivid illusion that the joint is being moved in the direction *opposite* to that in which the contracting muscle would be moving it.[2] So we have to assume that if the input from the spindles is quantitatively sufficient to have *any* kinaesthetic effect, that effect will be *kinaesthetically-untruthful*.[3]

The problem is: how is the fusimotor activation of flexor spindles, in the service of the automatic control of voluntary movement, prevented from making us feel that our will-to-flex a joint is causing it to extend? One answer might be that the kinaesthetically-untruthful effect is always quantitatively insignificant in relation to all the kinaesthetically-truthful effects described in Chapter VI. A second answer might be that there is interaction between the inputs, the kinaesthetically untruthful inputs being *gated out* by the truthful ones. A third might be that there is some central nervous process, possibly mediated by the 'corollary discharges' of corticospinal axon collaterals, which prevents the kinaesthetically-untruthful information from reaching the 'sensorium'. Of course the three answers are not necessarily comprehensive or mutually exclusive.

2. *Illusory perceptions aroused by vibrating the tendons of contracting muscles*

In the relative absence of detailed studies of vibration of finger tendons (Chapter VI) it seems permissible to think about experiments on vibrated elbows when wondering how active movements of the fingers are perceived. Neuropsychologists have long been interested in illusions for the light they can shed on normal perception.

Most of the experiments of the kinaesthetic effects of transverse percutaneous vibration have been done on biceps and triceps. But caution is necessary. The forearm is heavy, and automatic control and kinaesthesis at the elbow may well differ in detail from those of the fingers. Extension of the elbow is often effected by gravity, the triceps remaining relaxed while the flexors, brachialis and biceps, 'pay out' the movement. The actions of biceps and triceps are not limited to the humero-ulnar joint—both act also across the scapulo-

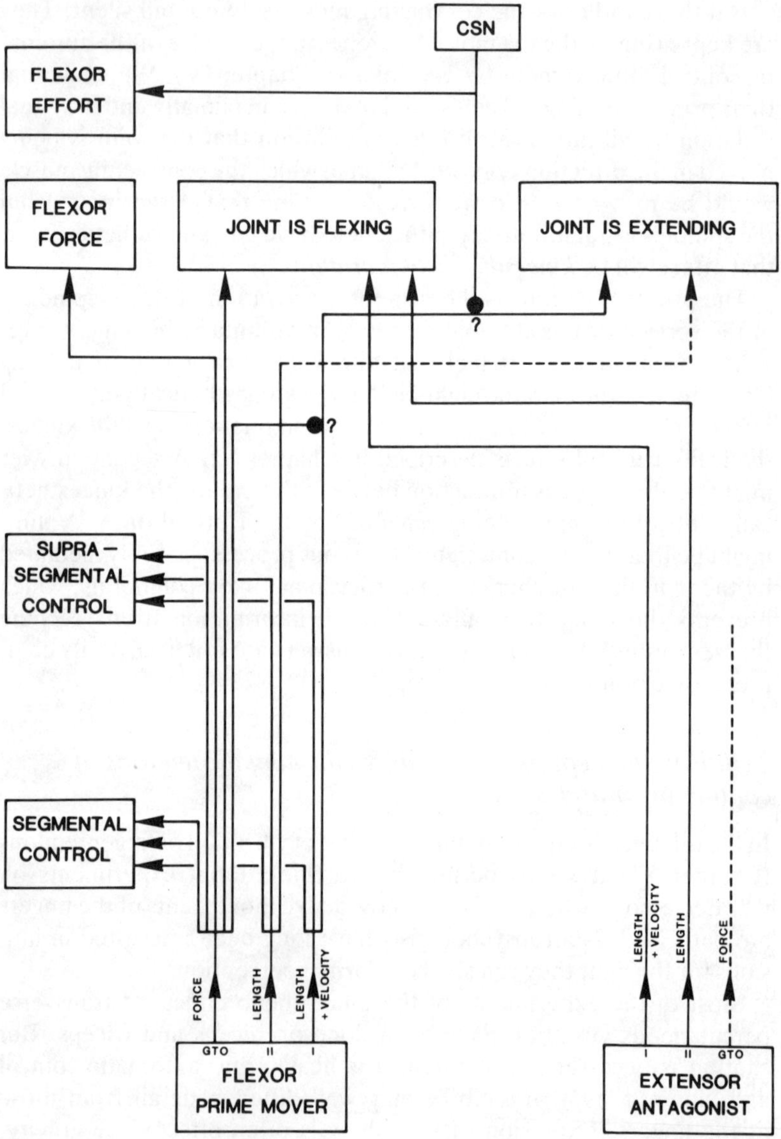

Fig. 11 Musculotendinous imputs aroused when biceps is acting as flexor prime mover. See text

humeral joint, and biceps is also the most powerful supinator of the forearm. But the illusions of flexion and extension of the elbow when their tendons are vibrated are always vivid. Let us take biceps as our model of a prime mover, with triceps as its antagonist, coupled mechanically, as in Fig. 7, but across the humero-ulnar joint.

In considering the perception of active flexion of the elbow it will be helpful to have a summary diagram of the musculotendinous inputs that will be affected when biceps contracts as a prime mover (Fig. 11). The inputs from the flexor prime mover and its extensor antagonist are labelled I (primary spindle endings, registering length-plus-velocity); II (secondary spindle endings, registering length and therefore joint angle), and GTO (Golgi tendon organs, registering force). The inputs from the prime mover are shown as distributed separately to the segmental control networks (Fig. 7), to suprasegmental long-loop control networks (Chapter IV), and to kinaesthetic and 'dynamaesthetic' networks (at the top of the diagram), to stress that the same signals could be 'read' separately, without any conflict between the 'requirements' of perception and motor control. Nobody should be misled by the absurd diagrammatic oversimplification of Fig. 11. The complexity of the real neuronal hardware must be very considerable. Suprasegmental pathways, represented by lines in the diagram, are in reality interrupted by 'relay nuclei'—collections of second-, third- and higher orders of neurones on which many centripetal and centrifugal impulses converge, their interactions depending on delicate balances of excitation and inhibition as much as on hard-wired connectivity.

Consider first an intriguing situation which does not correspond to either a passive or a voluntary movement. Vibration of its tendon readily excites a reflex contraction of biceps—the tonic vibration reflex. The prime mover is being driven by the input from its own spindle primaries, not by voluntary 'commands' from the brain. The driving is probably not wholly over the monosynaptic Ia network (Fig. 7), because the reflex starts gradually and may take several seconds to develop. The blindfolded subject's perception of the reflex movement is most efficiently communicated to the experimenter by getting him to track its course with the opposite arm. The trajectories of the vibrated and the tracking arm can be recorded with simple apparatus as in Figs. 12 and 13.

In this experiment there is no illusion that the elbow joint is *extending*. It is perceived as flexing; the *illusion* is that it is flexing more slowly than it really is (Fig. 12). Objectively, the recorded movement

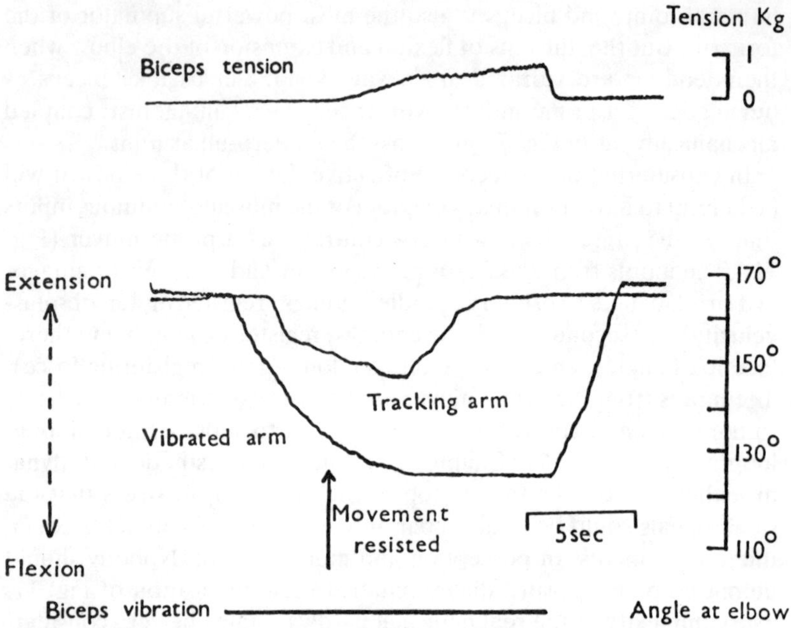

Fig. 12 Illusory perception of tonic vibration reflex. See text. (Goodwin *et al.*, *Brain* **95**, 705–48, 1972)

of the tracking arm is lagging behind that of the vibrated arm. The illusion can be interpreted, possibly over-simplistically, as the result of conflicts between the inputs illustrated in Fig. 11. Since the joint *is* flexing, the inputs signalling 'joint-is-flexing' will all be active in the same sense: the only kinaesthetically-untruthful input comes from the vibrated primaries of the prime mover, signalling 'joint-is-extending'. A balance is struck in favour of 'joint-is-flexing', but tilted towards 'joint-is-extending', so that the subject perceives the reflex flexion as slower and less ample than it really is.

In the experiment shown in Fig. 12, the reflex movement was arrested at the arrow without the subject's knowledge: his hand had been made insentient by anoxia so that he felt nothing when it was checked by the stop. Though the elbow joint was now stationary (the tonic vibration reflex having been suddenly made isometric instead of isotonic), the subject signalled his perception that the joint had moved into full extension. The surprised expression of a naive subject when he is allowed to see the final positions of his arms—the vibrated elbow flexed, the tracking elbow extended—never fails to amuse the onlookers.

A possible interpretation is that the balance of kinaesthetic inputs was shifted in favour of extension. When the flexion of the joint was arrested, this would have been registered by a sudden reduction of input from the antagonist's spindles and possibly by the joint itself. The massive signal from the prime mover's spindles, signalling 'joint-is-extending', would have been left in unopposed possession of the field.

In the isometric condition there was increasing flexor force (biceps tension, Fig. 12). This would have stimulated the prime mover's tendon organs; but if these are correctly represented in Fig. 11 as contributing to 'joint-is-flexing' (see p. 66), and as tending to block the spindles' access to the 'sensorium', their contribution was overwhelmed by the massive 'joint-is-extending' signal from its vibrated primaries.

If vibration is superimposed on a strong isometric voluntary flexion there is no illusion of extension. Strong voluntary contraction introduces a new element into the experiment: activity of corticospinal and other cerebrospinal neurones which not only drive the movement in α-γ-linkage but also discharge collaterally into all levels of the ascending somaesthetic pathways. There will also be strong excitation of the prime mover's tendon organs. It is tempting to speculate that powerful corollary discharges by corticospinal collaterals and maximal tendon organ inputs can block the pathway from the prime mover's primary endings to the 'joint-is-extending' part of the 'sensorium'. The kinaesthetically-untruthful input from the spindles would then be eliminated. These speculative possibilities are marked with filled circles and queries in Fig. 11. The blocking or 'gating' would have to be done at a level or levels that would not interfere with the access of the prime mover's primary endings to the automatic control loops, segmental (Figs. 7 and 9) and suprasegmental (Fig. 9).

Consider, next, the effect of vibration of biceps on *voluntary* flexion of the elbow (Fig. 13). The subject was voluntarily tracking a passive movement that was being imposed by the experimenter on the opposite elbow. The arms were horizontal, and the voluntary flexion and extension of the tracking forearm would both have been controlled by biceps, gravity acting as prime mover for the extension. The experiment on flexion (from 160° to 90°) is shown on the right half of the figure. For the first 10° of flexion the voluntary movement faithfully followed the passive one. Then when the vibration was applied, the

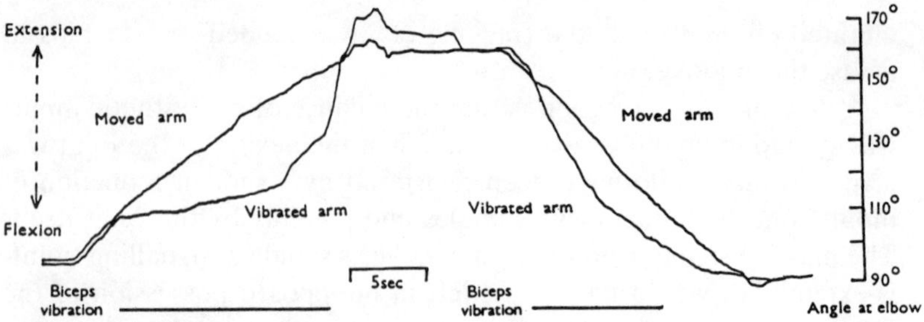

Fig. 13 Effect of vibration of prime mover on performance and perception of voluntary movement. See text. (Goodwin *et al.*, *Brain* **95,** 705–48, 1972)

voluntary movement was speeded up, running ahead of the movement it was supposed to be tracking. The subject, however, was unaware of the disparity between the trajectories of the 'target' and 'tracking' arms: he had the *illusion* that he was still tracking correctly.

On the left half of Fig. 13 the extension of the tracking arm (from 90° to 170°) was effected by gravity, controlled by graded voluntary relaxation of biceps. Vibration, which had speeded up the flexion, now slowed the extension of the 'tracking' arm so that it lagged behind the 'target' arm. Again the subject was unaware of his error: he had the *illusion* that the tracking was proceeding at the correct speed and that the 'tracking' arm was at every instant in its correct relation to the 'target' arm. When the vibration was withdrawn he quickly corrected the positional error.

Disturbance of the tracking *performance* by the vibration is not surprising. A tonic vibration reflex had been superimposed on the programmed voluntary contraction of biceps; the excessive Ia input disrupted the balanced α-γ- linked control descending from the brain. But what prevented the subject from perceiving his tracking error? Obviously, the illusory failure to perceive it would frustrate its correction by voluntary action.

Consider only the flexion movement. When successfully tracking it, the speed of the voluntarily-flexing elbow would have been correctly signalled to the 'sensorium' by all the kinaesthetic inputs of Fig. 11 except the prime mover's spindles. Hypothetically, either their kinaesthetically-untruthful input was quantitatively insufficient to disturb the correct perception of 'joint-is-flexing', or it was actively 'gated out' by the motor programme.

The programme itself should not have been affected by vibrating the voluntarily-flexing biceps. That programme would be controlled wholly by signals emanating from the opposite, passively-moved arm, and would be generating an α-γ-linked motor output to match the trajectory of the passively-flexed elbow.

When the voluntary flexion was speeded up by the vibration, the speeding-up would have been correctly signalled by all the kinaesthetically-truthful 'joint-is-flexing' inputs of Fig. 11. But suppose that the kinaesthetically-untruthful 'joint-is-extending' signalling by the vibrated spindles was massive enough to break through the 'gate'. If the resulting conflict between the truthful and untruthful inputs had been approximately balanced at the 'sensorium', the increased velocity of the voluntary tracking movement would have gone unperceived.

When biceps is exerting voluntary *isometric* force, and a tonic vibration reflex is superimposed on the voluntary contraction, there is an illusion, in some subjects but not in all, that the elbow is *flexing*.[4] The subjects were instructed to maintain a steady force. In the initial equilibrium the force would be generated by α-γ-driving by the corticospinal neurones. Hypothetically the access of the primaries to 'joint-is-extending' would be blocked. The tendon organs would be signalling 'flexor force' (Fig. 11, top left). The remaining inputs of Fig. 11, and the joint itself, would (again hypothetically) signal that the joint was stationary—any 'joint-is-flexing' signal from the tendon organs being balanced out by the other inputs. When the equilibrium was disturbed by vibrating the biceps tendon there was an increase in *force* of which the subjects were unaware (though they were aware of a lightening of *effort*). The 'extraneous' tonic vibration reflex had increased the force by interfering with α-γ-linkage in the motor control networks. Hypothetically there would have been a new kinaesthetic balance between the increased 'joint-is-extending' pressure from the vibrated spindle primaries and increasing 'joint-is-flexing' pressure from the tendon organs—tilted towards 'joint-is-flexing' in the subjects who experienced the illusion.

The negative illusion experienced by all subjects—failure to detect the increased output of isometric force—will be considered later (Fig. 15). Only if the subjects were allowed to see the dial of the strain-gauge did they correct the force, by a brief transient response at the onset and conclusion of vibration.

3. *Illusion of flexion in a stationary index finger*

It is time to return from speculations about the elbow to consider the results of some elegant experiments on the index finger.[5] Isometric voluntary flexor force exerted at the proximal interphalangeal joint of the index finger arouses the illusion that the joint has flexed. The illusory angular movement, as measured by the subject's reproduction of it by displacement of the opposite finger, is linearly related to the measured isometric force. In this experiment there is no interference with α-γ-control by an 'extraneous' Ia input. The perception is related to the *force achieved*, not to *effort exerted*. If the muscle is weakened by fatigue, the illusory perception is related to the diminished force exerted; the subject feels that his effort has been increased (see below, p. 71). This experiment seems to validate the contribution of tendon organs to 'joint-is-flexing' in Fig. 11. Hypothetically the side-effect of the fusimotor-activated Ia discharge ('joint-is-extending') would be 'gated-out' by the tendon organs and cortico-spinal neurones (Fig. 11). The remaining inputs of Fig. 11, signalling 'joint-is-stationary', would be quantitatively inadequate to balance out the preponderant kinaesthetic signal from the tendon organs. But so long as the joint is free to flex, even against large isotonic forces, there is no illusion: the subject's movements of his opposite finger show that the real angular displacements have been accurately perceived.

It is easy to forget, when concentrating on a single interphalangeal joint, that the greater the force exerted the more widely the activity spreads to include synergic and fixator muscles acting at other joints in the arm, even in subjects trained to relax as much as possible during the tests. Inputs from outlying areas of such widened receptive fields may well be making their contribution to the subject's awareness of the movement or force he is perceiving at the focus of attention and action.

Much of this section has been given over to the sort of beguiling but inconclusive speculation one can get involved in, but should not take too seriously unless it can be seen to point the way to a new experiment. There are neurones in the somaesthetic area of the cerebral cortex of monkeys which are 'kinaesthetically truthful' in that they fire impulses in response to angular movements of a joint in one direction and not in the other, whether the movements are active or passive (see Chapter IX). In experiments with microelectrodes it should be possible to discover whether their response to the unnatural, isolated *vibration* of a prime mover is 'kinaesthetically untruthful' with reference to the direction in which contraction of the prime mover would have moved the joint; and if so, whether this input can be 'gated' by artificial excitation of circumscribed populations of corticospinal neurones, Golgi tendon organs, or by any other inputs or combination of inputs (Fig. 11). The feasibility of more limited studies in man, using averaged sensory-evoked cortical potentials (SEP) instead of single neurones as in monkeys, has been shown by experiments in which the cutaneous SEP from the thumb was gated by passive movements of the thumb.[6]

4. *Corollary discharges arouse no kinaesthetic perceptions, whether movement has actually occurred or not*

Enough has been said about the probable role of corticospinal collaterals in the decoding of mixed and ambiguous signals—of length and velocity from the primary spindle endings, and of movement and contact by slowly-adapting receptors in the skin of the fingers. Possibly more than enough speculation has been risked on the gating and interaction of kinaesthetic inputs at loci that do not interfere with their operation in motor control. The question remains whether the corticospinal collaterals arouse any perceptions in their own right: the 'felt will' or 'sensation of innervation' of classical authors.

All the evidence shows that they arouse no perception of movements that have actually occurred or illusions of movements that have not. 'During recovery from a spinal anaesthetic,' notes Granit,[7] 'I myself ordered one of my legs, stretched out in bed, to be lifted. It felt dead and heavy and I was utterly unaware of the fact that it actually did move, until my toes bumped against the blanket and I had a blunt feeling of something like a thud.'

A more formal experiment is illustrated in Fig. 14.[8] The index finger of the left hand was anaesthetized by injection of local anaesthetic at its base. The blindfolded subject was asked to extend all the fingers of that hand, and then to imitate the resulting posture with the right hand. There was a striking failure to perceive that both interphalangeal joints of the left index had been fully extended—the proximal joint by the long muscle, extensor indicis, the distal by the intrinsic muscles, lumbricals and interossei. These latter must, at least partially, have escaped motor paralysis, though their afferent axons may have been selectively blocked. Certainly no corticospinal collateral action aroused any perception of the successful movement. One notes in passing that the sudden passive stretch of the long flexors by the extensor prime movers did not, on this occasion, arouse any kinaesthetic sensation.

In another experiment, deliberate advantage was taken of the first

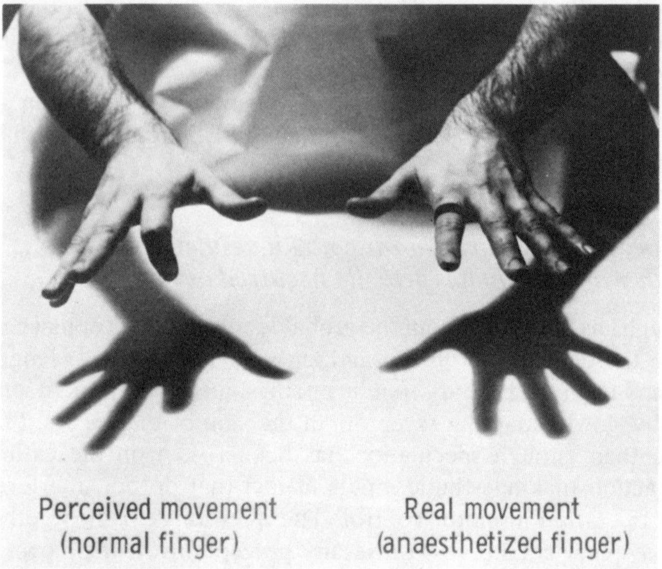

Fig. 14 Failure to perceive voluntary movement of an anaesthetic finger. See text. (Goodwin *et al.*, *Brain* **95,** 705–48, 1972)

few minutes of ischaemic-anoxic anaesthetization of the forearm and hand, when blockage of afferent precedes blocking of efferent nerve-conduction. The blindfolded subjects were asked to raise the index-finger into full extension at the metacarpophalangeal joint at half-minute intervals, and then to imitate the movement immediately with the opposite hand. As the time of complete paralysis drew near the subjects began to underestimate the extent of their movements. By the time the real movement had declined, in one experiment, from 40° to 30° the perception had declined to 5°; the three final movements were not perceived at all. If perceptions of movement had been aroused by kinaesthetic signals from corticospinal collaterals, there should have been less decline, or even the illusion that there had been no decline.

No illusion of movement is felt when a subject attempts to move paralyzed muscles. Under local curarization of the forearm and hand, people knew that they could not move 'and were unanimous in their denials of any sensations of movement accompanying their attempts. When movements were achieved by muscle groups which were weakened but not paralyzed they were perceived with apparently normal accuracy.' There is no doubt that in these circumstances the brain is despatching messages to the spinal cord, and the spinal cord towards the muscles: the motor nerve impulses can be recorded from the peripheral nerves, proximal to a nerve-block or to neuromuscular junctions blocked by curare.[9]

5. *Perception of voluntary movements of phantom limbs is not due to corollary discharges*

In the classical controversies between 'inflow' and 'outflow' theories of the perception of movement, phantom limbs played a prominent part. Interest in phantom limbs goes back at least as far as Bell,[10] who took the perceptions of lost limbs as further proof of the existence of the muscular sense. 'Urging a patient to move who has lost his limb, I have seen him catch at the limb to guard it, forgetful that it was removed; and long after his loss, he experiences a sensation not only as if the limb remained, but as if it were placed or hanging in a particular position or posture. I have asked a patient—"Where do you feel your arm now?" and he has said, "I feel it as if it lay across my breast," or that it is "lying by my side." It seems also to change with the change of posture of the body. These are additional proofs of a muscular sense; that there is an internal sensibility corresponding with the changing condition of the muscles; and that as the sensations of an organ of sense remain after the destruction of the outward organ, so here a deceptious sensibility to the condition of the muscles, as well as to the condition of the skin, is felt after removal of the limb.' Bell had distinctly adumbrated the 'inflow' and 'outflow' theories in 1833 (see p. 41) and declared himself firmly for 'inflow'.

In his classic textbook, Gowers[11] wrote eclectically: 'Our knowledge of active states of the muscle is due, at least in some measure, to the effects on consciousness of the activity of the nerve-structures causing the movement ... But to its effects must be added that of the impulses from the muscles.' One of the 'proofs' he offered in support of the first statement was that 'after amputation of a limb, a person who makes an effort to move the lost part seems to feel as if he did move it.'

The 'proof' has been invalidated by a definitive study of about 300 patients with phantom limbs.[12] There is an important distinction between movements of the phantom as a whole, moving with the real joint that moves the stump, and movements of phantom joints (*within* the phantom). The latter are always associated with movements of muscles within the stump, and are abolished when these muscles are denervated. Presumably such contractions would represent surviving fragments of the complex totality that would have formed part of a voluntary movement. Movements of phantom joints are always limited and poorly graded. Awareness of movements of the fingers aroused by contractions of remnants of proximal muscles may recall some previous comments on perception and performance in chronic deafferentation (p. 21) and on the possible contribution of physiological irradiation of activity to proximal muscles when the perceptions of a single digital joint are being examined (p. 66).

Disposal of the claim that voluntary movements of a phantom prove the existence of a 'sensation of innervation' does not deny the existence of the phantom. Bell's 'deceptious sensibility' remaining 'after the destruction of the outward organ' later attained greater sophistication as Head's 'body image'.[13] Neither author would perhaps have gone as far as Admiral Lord Nelson, who is said to have regarded his phantom arm as evidence for the immortality of the soul.[14] He did not know of cases in which the phantom was lost after a lesion of the parietal cortex.

6. *Perceptions of force and effort*

But, if corticospinal collaterals arouse no perception of *movement*, they are certainly involved in our perceptions of *effort*. Perceptions of efforts exerted and forces achieved are very closely related in normal people, but are distinguishable in clearly-designed experiments in which the instructions given to the subject are carefully worded.

Left to themselves, subjects are apt to attend to subjective effort rather than to objective force. But if appropriately instructed they can attend to either.[15] A subject maintaining a prescribed isometric flexor force at the elbow is asked to keep his effort constant. Then during a period of vibration of biceps a tonic vibration reflex is simply grafted on to the voluntary contraction. In another experiment he is asked to keep the tension constant. When the biceps is vibrated the rise in tension is corrected, and when tension falls when vibration is withdrawn, the fall in tension is corrected. When biceps motoneurones are reciprocally inhibited by vibration of triceps in 'constant effort' and 'constant tension' experiments, the subject makes the appropriate adjustments (Fig. 15).

Thus perceptions of force and effort can be dissociated by the subject's will. The experimenter for his part can measure objectively

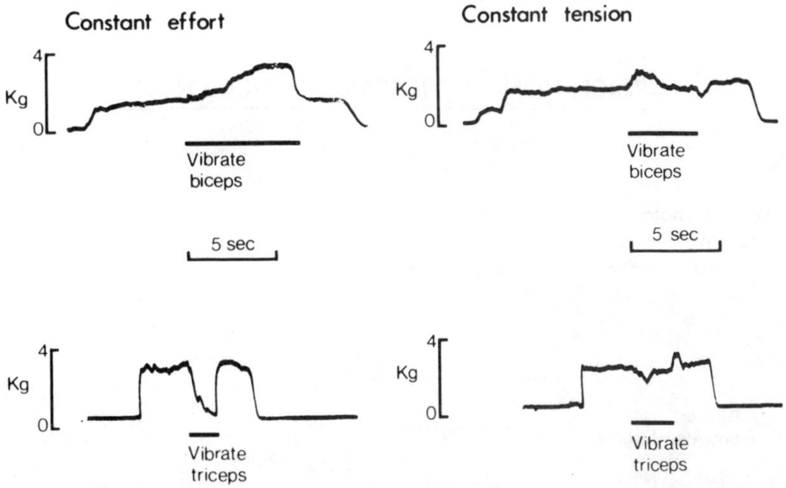

Fig. 15 Subject instructed to exert constant effort or maintain constant force. See text. (McCloskey, *Physiol. Rev.* **58,** 763–820 (Fig. 4), 1978)

the subject's output of isotonic or isometric force; and he can also make quantitative estimates of perceived weights or tensions in experiments in which the subject has to match these perceptions by operating a similar manipulandum with the opposite hand. For example, the effort required to lift a weight by pressing a lever with the distal phalanx of the thumb can be matched by the opposite thumb pressing on a similar lever, the weight being adjusted until the sensations on the two sides appear equal. The subject's cognitive 'set'—for 'force' or 'effort'—depends critically on the wording of the instructions.

The contributions of peripheral inputs and corticospinal collaterals (and other central mechanisms) can be examined in situations in which their normal balance has been upset experimentally or by disease (Fig. 16). When the prime-mover muscles have been weakened by fatigue, partial curarization, reciprocal inhibition of their skeletomotoneurones by vibration of the antagonist, cerebellar lesions, or paretic strokes without accompanying sensory disorder, the subject feels sensations of 'heaviness' and of unnaturally large effort. This is quantified by the increased weight demanded by the subject to match the effort with the opposite thumb. The input from the tendon organs of the weakened muscles can be assumed to supply accurate measurements of the forces exerted. Weakened muscles or inhibited skeletomotoneurones require stronger α-γ-driving by the corticospinal

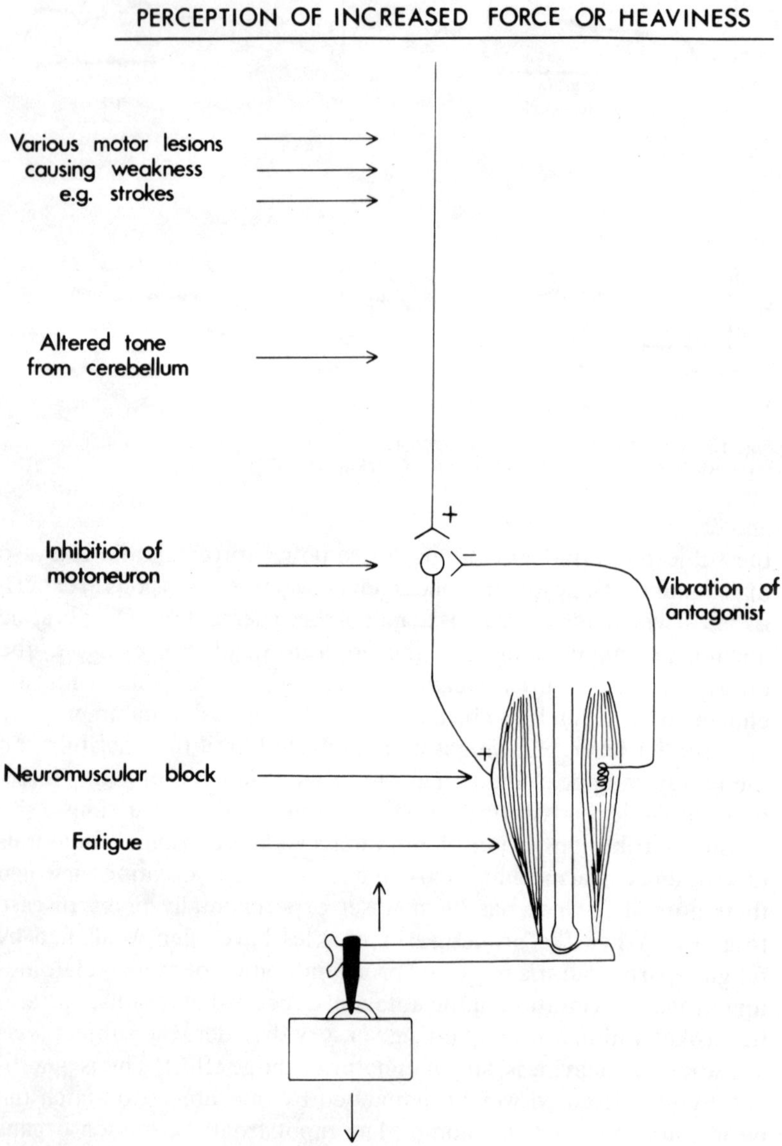

Fig. 16 Factors increasing perceived heaviness. (McCloskey, *Physiol. Rev.* **58,** 763–820 (Fig. 5), 1978)

neurones, entailing stronger collateral actions on the proprioceptive and somaesthetic pathways which are presumed to arouse the classical 'sense of effort' or 'felt will'.

In cases of pure motor stroke, the 'sense of effort' is increased in proportion to the weakness. It can be lessened by vibration of the weak muscles. It is a reasonable hypothesis that the increased 'sense of effort' is due to the intense motor activity of a surviving but diminished population of corticospinal axons and their collaterals. This activity could be reduced if the voluntary movement was aided by a tonic vibration reflex. The hypothesis expects that total paralysis, resulting from the total absence of corticospinal neuronal activity, will be unaccompanied by any 'sense of effort'.

Mach, who, like Helmholtz, believed in 'sensations of innervation', described his own experience in 1886.[16] 'I was in a railway train, when I suddenly observed, with no consciousness of anything else being wrong, that my right arm and leg were paralyzed; the paralysis was intermittent, so that from time to time I was able to move again in an apparently normal way. After some hours it became continuous and permanent . . . I can only describe my condition during the period of complete paralysis by saying that when I formed the intention of moving my limbs I felt no effort, but that it was absolutely impossible for me to bring my will to the point of executing the movement. On the other hand, during the phases of imperfect paralysis, and during the period of convalescence, my arm and leg seemed to me enormous burdens which I could only lift with the greatest effort . . . The paralyzed limbs retained their sensibility completely . . . and thus I was enabled to be aware of their positions and of their passive movements.'

Mach's self-observations have been replicated in a recent report of cases of pure motor hemiplegia.[17] The attempts of one patient to extend her fingers went through three stages. While still unable to extend her fingers she had no feeling of heaviness. In the second stage there was 'a feeling of immense heaviness without any movement.' On the basis of this she predicted, correctly, that the movement would return. In the third stage the sense of heaviness diminished as the movement reappeared and became stronger.

In the earliest stages of his own recovery Brodal[18] noted: 'Subjectively it was felt as if the muscle was unwilling to contract, and as if there was a resistance which could be overcome by very strong voluntary innervation. The greater the degree of paresis of such a muscle, the greater was the mental effort needed to make it contract and to oppose voluntarily even a very weak counter-force. On the other hand only a slight mental effort was needed to bring about a fairly good contraction of a muscle able to work with about half or a little less of its full force ... The expenditure of this mental energy is very exhausting, a fact of some importance in physiotherapeutic treatment.'

The relation between 'sense of effort', perception of movement and perception of force achieved has been examined in normal people lifting weights by pressing levers with the terminal phalanges of their thumbs. A weight that is too heavy to be lifted cannot be discriminated against another weight. The results suggested to the experimenters[19] that the peripheral signals that give the subjects their 'knowledge of results' need only be relatively crude—'as if the peripheral signal is simply an "event marker" to indicate to the central nervous system the point in a ramp of centrally generated command at which the command has succeeded.' The instant of *lifting* would have to be signalled by the start of the *movement*.

Yet again the deafferented patient re-enters.[20] Recall that with visual guidance he could learn to generate three levels of force, adjusting these to the visual display of his force output during the learning of the task (Fig. 5). His adjustments of motor outflow, whether in isometric or isotonic experiments, were remembered (programmed) and could then be run off without vision until they disintegrated for lack of feedback from the hand. He could not judge weights because he did not know whether he had lifted them.

The next-best experiment was to get him to match isotonic forces. The distal phalanx of each thumb was resting against a lever. Each lever, when moved through 5° by the interphalangeal joint of the thumb, switched on an illuminated signal. The isotonic force required to move the 'target' level was set by the experimenter. The isotonic force that opposed the movement of the 'matching' thumb was varied by the experimenter, who asked whether the matching force was larger or smaller than the target force. The matching force was varied until the patient judged that the target and matching torques were equal. A new target force was then set up and the matching repeated. The plot of target torques against matching torques was linear. The experiment could not discrminate between the possible loci of the patient's apparently successful 'sense of effort'—upstream or downstream of the corticospinal neurones.

The patient's strategy, however, raised an intriguing question about the significance of the result. It seemed possible that the task he had successfully accomplished was to send the *same* motor command to each thumb. He said that he made the final match in each sequence by 'trying to press with the same force with both thumbs'. He estimated which thumb had the larger load by comparing the time each lever took to move through the 5° and switch on its light. 'Using this technique, he did not need any knowledge of the absolute size of the voluntary motor command, or even the ability to compare the output to each thumb.' But if the experiment could not prove that he had this latter ability, neither could it prove that he had not.

CHAPTER VIII

Active touch

Sherrington seems to have been the first to use the term 'Active Touch',[1] but the idea of it was clearly stated by Charles Bell in his Bridgewater Treatise on the Hand.[2] 'Without a sense of muscular action or a consciousness of the degree of effort made, the proper sense of touch could hardly be an inlet to knowledge at all . . . The property of the hand of ascertaining the distance, the size, the weight, the form, the hardness and softness, the roughness or smoothness of objects results from the combined perception—through the sensibility of the proper organ of touch and the motion of the arm, hand and fingers. But the motion of the fingers is especially necessary to the sense of touch . . . embracing the object, and feeling it on all its surfaces; sensible to its solidity and to its resistance when grasped; moving round it and gliding over its surface, and, therefore, feeling every asperity.'

1. *Tactual and muscular impressions*

In his textbook article on Cutaneous Sensations, Sherrington[3] gave greater precision to these ideas. He noted that for judgement of the size and shape of objects by touch, it is necessary to excite 'a certain number of tactual sensations of definite intensities and imbued with "local signs". These component sensations are combined, either (1) simultaneously or (2) successively as well as simultaneously. The successive component sensations can be acquired by *actively moving the tactual surface* (his italics), to increase or renew the simultaneously given groups of sensations, and in so doing the muscular sense is combined with the tactual in obtaining criteria for judging the shape and size of the object by the distances moved.' In considering the relations between movement and tactile exploration, he identified two extremes: (1) 'the same tactual point' being moved over the explored object: the same touch-sense related to 'a series of "muscular" sensations differing in intensity: "active touch"; (2) An object is allowed to move over a series of tactual points, the tactual surface being held stationary, i.e. one muscular sensation is combined with a series of tactual; "passive touch".' The superiority of the moving stimuli elicited by active exploration is reflected in a property of passive touch: successive stimuli are more easily interpreted than simultaneous—'It is easier to read letters as they are traced on the skin than letters already fully shaped and pressed upon the skin'.

As in the case of active movement, illusions have highlighted the problems about how the normal processes of tactile discrimination interact with the perceptions of position and movement. One of these illusions goes back to Aristotle. If the index and middle fingers are crossed and a pea is placed between them, 'the pea is "felt double", as though it were two peas. The illusion is strengthened by moving the surfaces of contact. The two sides of the fingers in use are in their ordinary position invariably turned away from each other. Experience has taught that they usually touch two different objects, and this association may be, in the unpractised person, sufficient to override the corrective perception furnished by the muscular sense.' Other illusions of passive touch are related to movement. 'The points of a pair of compasses, when moved with a sufficient speed over a length of skin, appear to get wider apart as they pass from skin less endowed with local sign to skin more endowed, e.g. from ear to lip (Weber) . . . If moved with uniform speed, they appear to travel faster as they pass over the more perceptive region (Vierordt).' Such illusions invite formal replication along lines running from volar surface of forearm to finger-tips, with comparison between the perceptions aroused by points moved passively along the stationary skin and skin moved actively across stationary points. But such a 'planar' approach would have limited interest. For the perceptions of everyday life are based on four-dimensional explorations of three-dimensional surfaces and of smaller solid objects that can be taken into the hand (stereognosis). Active touch discriminates size, shape, indentibility and surface texture by compound perceptions—kinaesthetic, of force, speed and direction of finger movements; tactual, of vertical deformation, horizontal drag, and vibrations of different frequencies generated by the sliding of glabrous skin over variously-textured surfaces; and thermal, of temperature gradients across the skin. The totality must be referred to the brain-models of familiar surfaces and objects, built up during the life of the individual, or fed into the learning processes by which the properties of hitherto-unfamiliar objects are added to the repertory.

2. *The somaesthetic macula*

The glabrous skin of the finger-tips, hairless, thrown into papillary ridges (finger-prints) kept supple by sweat, and profusely innervated by fast-conducting axons, is the somaesthetic macula. Microneurography has confirmed and extended in man the properties of the mechanoreceptors first discovered in the glabrous skin of cats and monkeys.[4] Like the muscle receptors, some respond dynamically during transient deformations (Fast-adapting, FA; about 55%); others statically, throughout sustained deformations (slowly-adapting, SA; about 45%). There is a second subdivision into two classes. FAI and SAI have small, sharply-circumscribed receptive fields, 2 to 8 mm in diameter, spanning four to ten papillary ridges and containing several spots of lowest threshold. They are most densely distributed in the distal halves of the distal phalanges (FAI, $140\,cm^{-2}$; SAI, $70\,cm^{-2}$) where there is considerable overlap, so that a 'point' would overlie parts of 20 FAI fields. They are specially responsive to contours, discharging more impulses when their receptive fields are crossed by an edge than when the receptive field is indented as a whole. These properties associate FAI and SAI with the two receptor-types which lie superficially in the dermis, the Meissner's corpuscles and Merkel's discs.

When tested by vertical indentation of the skin, FAII and SAII have wide receptive fields with indistinct borders, covering a whole finger or a large part of the palm. They account for about 30% of the mechanoreceptors of glabrous skin. Their properties relate them to structures lying deep in the dermis. FAII are the Pacinian corpuscles. These are extremely sensitive to vibration: frequencies of 100 to 300 Hz are faithfully followed by the axon, even when the deformation of the skin is only 1 μm in amplitude. They would be readily stimulated by the vibrations set up when a finger-tip is moved at different speeds across rougher or smoother surfaces. SAII receptors have been mentioned already in connection with movement of the finger-joints. They respond to horizontal traction of the skin rather than to vertical indentation. Some of them show directional sensitivity, firing most impulses when the traction is in a specific direction. Many discharge statically in relation to joint angle. They correspond to Ruffini endings, whose relation to bundles of collagen fibres running in the deep layers of the dermis would fit them for the role of 'exoskeletal' finger-joint receptors. Similar histological structures are found in joint capsules and interosseous membranes. But all four classes of mechanoreceptors are excited by movements: all FAII units, 57% of FAI units and 66% of SAI units give some response to movements in which no external object has touched the skin.

These mechanoreceptors can elicit movements as well as arousing sensations.[5] Vibration of the pulp of thumb or index always evokes a reflex flexion of the digit in subjects exerting a weak flexor effort. In five out of 20 subjects this exteroceptive reflex occurred in digits that were passive and relaxed. Control experiments distinguished the large contribution of the glabrous mechanoreceptors from the smaller contribution of tendon vibration (the familiar tonic vibration reflex). 'The present results indicate that afferent signals from such receptors, by a feedback mechanism, also might influence the motor performance of the exploring fingers'. There is evidence (to be presented below) that minimal slippage of an object held in the precision grip of thumb and index elicits a corrective increase in grip force at 'long loop' latency.

3. *Tactual abilities of the blind*

Passive and active touch must reach their highest development in blind people. Passive touch is exploited to the full in the Optacon, a reading device for the blind which is only made possible by the spatial discriminating ability of the distal pulp of the index finger.[6] (Drastically less good results are obtained from the proximal half of the distal phalanx.) The pulp rests on an array of punctate vibratory stimulators, arranged in six columns of 24 rows within an area about 25 × 13 mm. The array is contoured to fit the finger-tip. Each point when activated vibrates at 230 Hz. A hand-held camera about the size of a pocket knife contains a corresponding array of 6 × 24 photosensitive elements. As the reader moves the camera along a line of print the patterns of light and dark are reproduced on the tactile array. Each letter is bounded by 1.1 × 2 cm on the finger-tip. Outlined by the appropriate configuration of vibrated points, the letter moves as a whole from right to left across the pulp as the reader moves the camera from left to right along the line of print. In the first day-and-a-half of a nine-day course, blind people can learn the *tactual* skills which enable them to read the alphabet; for the rest of the course they can give increasing attention to cultivation of the *linguistic* skills. Sixty words a minute, or better, can be read by many blind people.

Active touch is evidently important for the reading of tactual maps by blind people.[7] In recognizing the areal and linear symbols for road,

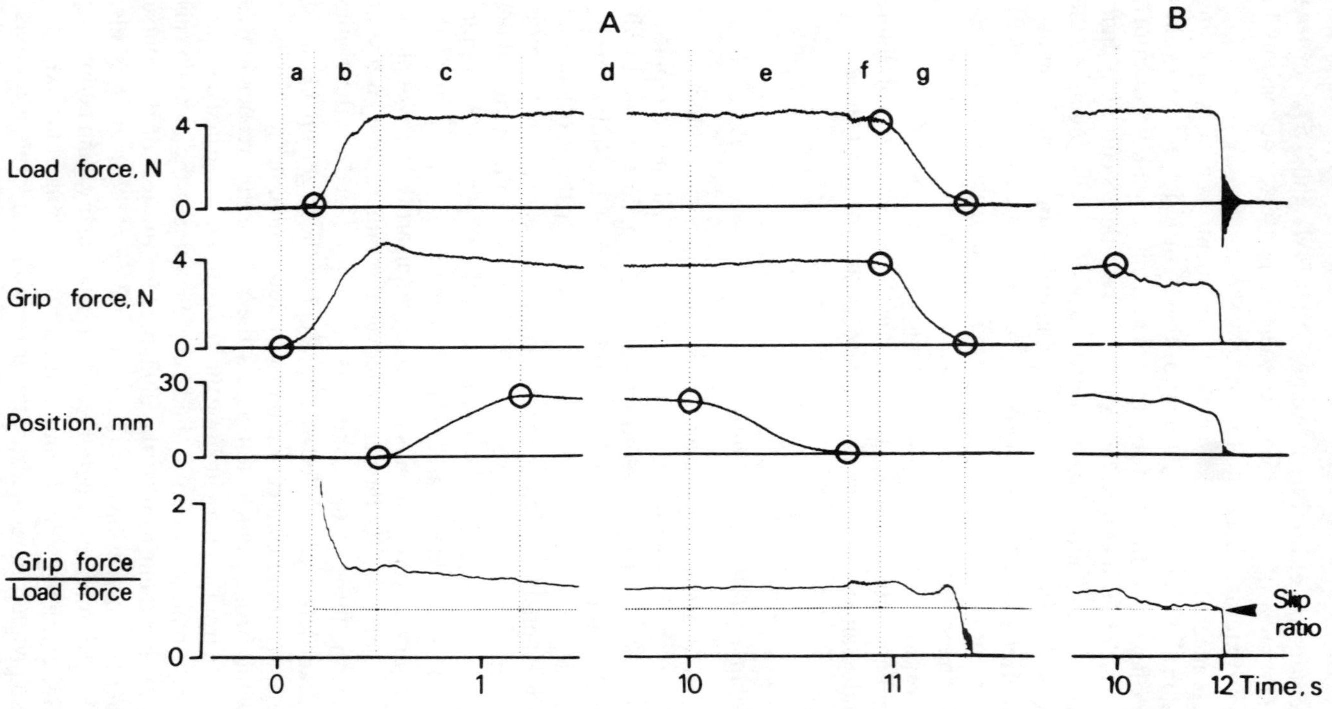

Fig. 17 Measurement of load force and grip force when lifting an object. See text. (Johansson and Westling, *Exp. Brain Res.* **56,** 550–64, 1984)

bus-stop, park, etc., 'the most successful subjects were those who made extensive though very fine exploratory finger movements.' This must also be the practice of the most expert readers of Braille. But the consummate achievement of active touch was the use made of it by the blind, deaf and indomitable Helen Keller, who received speech by placing the digit-tips of one hand on the moving lips and vibrating larynx of the person who was talking to her, while 'holding hands' with the other.

4. *Experiments on active touch*

In the face of such complexities, tactual and kinaesthetic, the problem of isolating sub-units of active touch for experimental analysis clearly calls for considerable ingenuity. Work at Umeå University on how we handle objects of different degrees of slipperiness is a model of what can be done.[8] The subjects' task was to lift off the table, with thumb and index, a manipulandum 30 mm wide and weighing 400 g, with interchangeable surfaces of silk, suède or sandpaper. The manipulandum contained transducers for measuring the vertical lifting force, the grip force, and the vertical movement; an accelerometer detected any slippage between the glabrous skin and the manipulandum. Fig. 17A shows the sequence of events determined by the brain programme. At time 0 the subject starts to grip the manipulandum. At the beginning of the loading period (b) the lifting elbow flexor force begins isometrically; the two forces reach maxima at the beginning of period (c), when the isotonic lifting movement (elbow flexion) begins. The required lift is completed at the beginning of the static period (d), when the isotonic lifting force again becomes isometric. At the start of (e) the subject begins to replace the manipulandum on the table, which it reaches at the end of (e). After a delay (f) the lifting and grip forces fall to zero.

The automatic adjustment of the isometric grip force by the programme is the nub of the experiment. Too little force and the manipulandum will slip through the precision grip of thumb and index. Unnecessary force would crush a brittle object, and, in the everyday situation, would hamper the delicate superimposed exploratory movements of active touch. The ratio grip force/load force has a value, the *slip ratio*, at which the manipulandum will be dropped. This value is measured in Fig. 17B, starting from the same force levels as those recorded in the static period (d): the subject slowly weakens the grip (note the record of grip force) until the slip ratio is reached. This

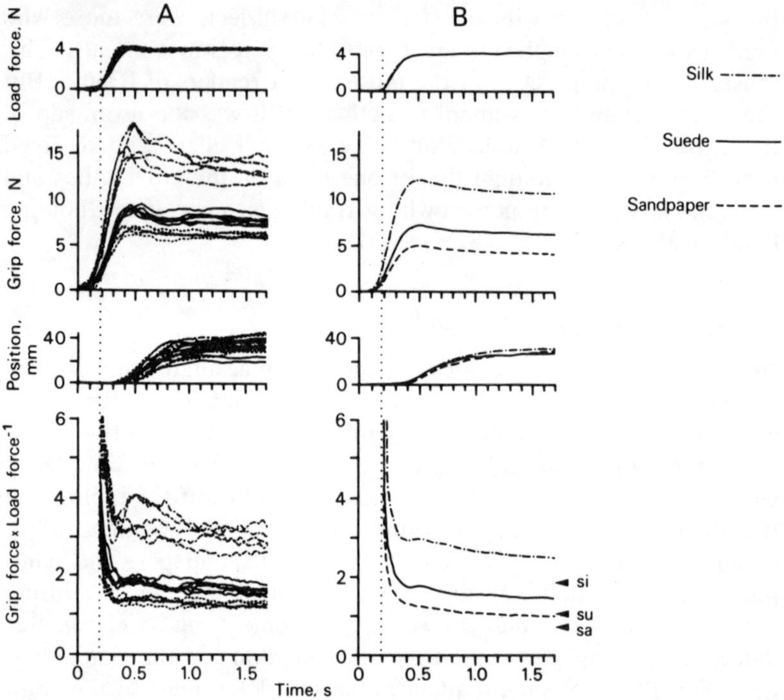

Fig. 18 Measurement of grip forces exerted in lifting objects of different degrees of slipperiness. See text. (Johansson and Westling, *Exp. Brain Res.* **56,** 550–64, 1984)

ratio is marked in Fig. 17A by the horizontal line. The difference between the grip force/load force ratio (coordinated automatically by the subject's brain) and the slip ratio is the *safety margin*. It is not very large.

Fig. 18 shows what happens when the surface of the manipulandum is silk (most slippery), suède (less slippery) or sandpaper (least slippery). The time-scale spans periods (a), (b) and (c) of Fig. 17. Fig. 18A shows superimposed records of sixteen trials by one subject, and Fig. 18B shows averaged records of 120 trials by nine subjects. The trajectory of the lifting movement ('position', Fig. 16) and the rate of development of the load force are unaffected by the different frictional conditions. Only the grip force changes. It rises most steeply, and to the highest value, for silk, least steeply and to the lowest value for sandpaper, with suède in between. The slip ratios for silk, suède and sandpaper are shown at the bottom of Fig. 18B: the safety margin is larger for silk, smaller and about equal for suède and sand-

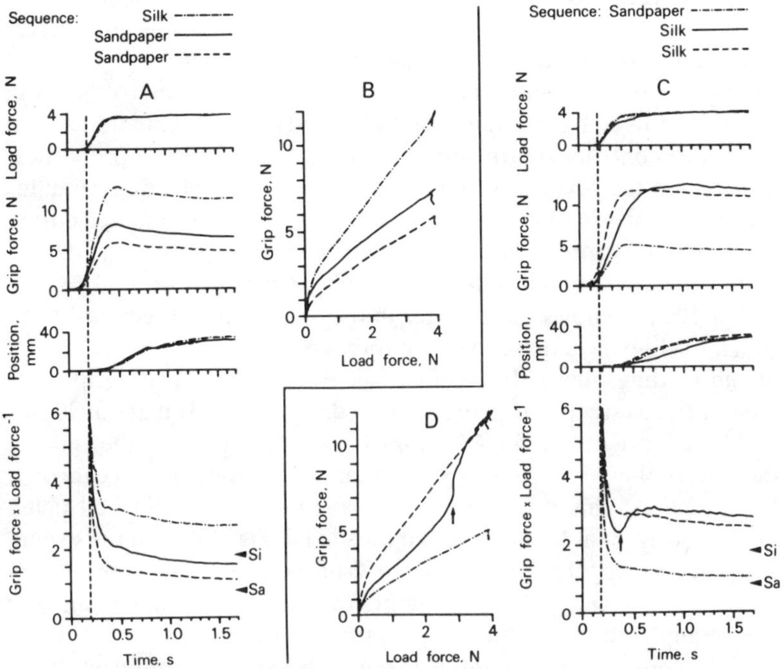

Fig. 19 The process of adaptation of grip force to different degrees of slipperiness. See text. (Johansson and Westling, *Exp. Brain Res.* **56,** 550–64, 1984)

paper. For each of the surfaces, a different ratio of grip force/load force is programmed automatically by the brain. Further analysis showed that the programmed adjustment of forces to frictions 'included a setting of a constant ratio between the rate of change of the two forces, rather than a regulation of the force ratio.'

The process of adaptation to a different surface (the weight of the manipulandum remaining the same) is illustrated in Fig. 19. Fig. 19A, B shows the changes in the 'safe' direction—from slippery to rough—in three consecutive trials, silk–sandpaper–sandpaper, averaged from 162 trials in nine subjects. Again, the development of lifting force and the trajectory of lifting are unaffected. During the pre-load period the development of grip force is the same in the silk and first sandpaper trials. During the following loading period the grip force rises less steeply in the first sandpaper trial (full line) than in the silk trial (dot-and-dash line), and to a lower final force, but the adaptation is not complete: in the second sandpaper trial (dashed line) the safety margin is reduced further. The curves deviate at the beginning of the loading phase, about 100 ms after the onset of the grip force. The change in the surface must have been detected at its first contact with glabrous skin and while there was only a minimal drag from the lifting force. The stimulus was presumed to be a minimal frictional slip, since control experiments showed that the frictional properties, rather than the textures of the surfaces, were the adequate stimulus for changes in grip force. The programme was partially updated during the first 100 ms (pre-load phase) of the first sandpaper trial, and completely updated before or during the first 100 ms of the second.

Fig. 19C, D shows the changes in the 'dangerous' direction—from rough to slippery. Again, the grip forces began to deviate at the start of the loading phase. In the first silk trial (full line) the grip force did not rise as steeply as in the second (dashed line). But about 200 ms after the start of the load (lifting) force, and at about the time of the start of the lift, there was a secondary rise of the grip force (arrow, Fig. 19C). At this point, the grip force/load force ratio had fallen 'dangerously' near to the slip ratio for silk (si, Fig. 19C). In the second silk trial the grip force rose steeply to its proper level, and no secondary correction was needed: the updating of the programme had already been completed during the previous trial with silk.

The secondary adjustments in the first silk trials were often preceded by small slips, detected by the accelerometer. Such adjustments were also seen during the static phase, when the safety margin was 'dangerously' small. The latency of these responses was 60 to 80 ms after the slips. The slips were rarely noticed by the subjects, and the responses were evidently automatic, not voluntary. These responses, and those evoked during the pre-load phase, are possibly mediated by a 'long loop' pathway from the glabrous receptors to the motor cortex (Chapter IX).

Anaesthetizing the distal phalanges of thumb and index (or the whole of both digits) highlighted the contribution of the glabrous skin receptors. The first lifts were successful with suède or sandpaper but not with silk: the digits simply slid up the stationary manipulandum without evoking any reflex response. A new force balance was then established by voluntary control, and its programme remained stable in subsequent trials. The pre-load phase was lengthened to about 500 ms, and at the start of the loading phase the grip force was about five times larger than normal. But the delay phase between the terminal contact of the manipulandum with the table and the switching-off of the load and grip forces (Fig. 17A, F) was unchanged (80 ms).

In the absence of responsive glabrous receptors the switching-off would have been by muscular reflexes (presumably suprasegmental). It seems possible that the delay period (f) was determined by muscular reflexes in all the experiments, even in those in which the glabrous skin receptors were intact.

The programme controlling the grip on the anaesthetized side was also accessible to inputs from the opposite side. The anaesthetized thumb and index lifted only the sandpaper-surfaced manipulandum, the unanaesthetized lifted silk-, suède- or sandpaper-surfaced manipulanda. The more slippery the object on the normal side, the higher was the grip force/load force ratio on the numbed side. So if, in everyday life, both hands are informed, the disengaged hand stands ready to intervene.

The experimenters have stressed the advantages of a preprogrammed coordination of grip and load forces, adjusted to the prevailing surface structure and rapidly readjustable whenever the structure is changed. The time-lags involved in *continuous* cutaneous-closed-loop monitoring are avoided by the alternative of *intermittent* resetting of an open-loop operation. Economical control of grip force (small safety margin) disengages a larger fraction of the motor apparatus for the exploratory manipulations thus underwritten by the non-slip programme in the service of active touch. Their suggestion that 'the proposed memory-based design would be of particular value during quick handling of fragile objects, e.g. while picking berries' is a pleasant reminder of many happy summer holidays in Scandinavia.

5. *Clinical aspects of active touch*

We take for granted our everyday abilities in tactile discrimination and manipulation. Everyone has felt the clumsiness of fingers numbed by cold, but there are harder lessons to be learned by patients suffering from injuries to the hand. Even in areas in which motor power is preserved, the symptoms of loss of cutaneous innervation are 'motor'—difficulty in gripping, picking up tiny screws, and so forth. Neurophysiology and neuropsychology have much to learn from the clinical experience of hand surgeons, notably from Sterling Bunnell and his pupils.

Previous work on 'repair' of cut nerves by end-to-end suture, or by insertion of free nerve grafts to bridge the gap left by an injury, took insufficient account of the *quality* of any cutaneous reinnervation that was achieved. H. J. Seddon distinguished between 'academic' and 'functional' recovery.[9] Moberg remarks: 'Children's hands are the only ones in which I really believe that the words *restoration of function* can be used.'[10] In adults, one year after operation, the only sensations aroused by touch stimuli are paraesthesiae. After three years these unpleasant sensations have become less, and some useful sensation for gross grip has been restored. After five years there is further improvement in sensation for gross grip. There is 'protective sensation', very important in that it enables reflex and voluntary withdrawal from injury. But normal use of the fingers remains impossible except in a few fortunate cases. In these there has been some degree of restoration of 'tactile gnosis'—due to successful regeneration of some of the fast-conducting afferent axons, and detected, not by pinprick and cotton wool, but by two-point discrimination and the ability to pick up small objects. It appears that 'tactile gnosis' is never fully restored after nerve suture in adults. The two-point threshold on the fingers may be only 6 to 15 mm, instead of the 3 to 5 mm accepted as normal by neurologists.

Such disappointing results after such long delays, and in spite of the introduction of microsurgery, have encouraged the alternative approach of transferring a surviving area of normally-innervated glabrous skin to a more strategic position in the hand. In one case, an area of thumb skin, which retained normal tactile gnosis from dorsal branches of the radial nerve, was moved forward on to the volar surface of the thumb. A 7 cm gap in the median nerve was bridged by insertion of a free nerve graft taken from the useless ulnar nerve. The result: the area of skin denervated by the ulnar nerve injury remained unprotected and liable to ulceration. The area supplied by the repaired median nerve recovered 'protective sensibility'. The normally-innervated skin on the tip of the thumb, 'no bigger than a postage-stamp', led the hand in all its gripping functions—a much reduced somaesthetic macula but enormously better than none at all. 'The follow-up of these patients, however, shows that not in a single instance, even after a number of years, did tactile gnosis expand beyond the original border of the flap . . . some small nerve branches form free axons which grow over in the adjacent skin area and others form microneuromata close to the border line of the flap. An examination with a pin therefore will necessarily give a response peripheral to the border. It is important not to misjudge this response as an increase in the area of useful sensibility but to understand that this paraesthesia is a minus factor not without importance. In some instances, the paraesthesia may even prevent the patient from using the digit. As it is impossible to avoid this minus factor, the procedure should not be used in patients incapable of overcoming these slowly-diminishing difficulties.'

In another patient, a heavy labourer, a deep lesion in the palm of the hand had healed with extensive scarring after infection.[11] The injury had severed all branches of the median nerve, the tendon of flexor pollicis longus, and the tendons of flexors digitorum superficialis and profundus of the index finger. The long flexors of the other fingers and the nerves to the thenar and hypothenar muscles were intact. All volar sensibility was lost except that of the little finger and the ulnar half of the ring finger. The skin of the thumb remained innervated by dorsal radial nerve-branches which spread unusually far volarwards from the dorsolateral aspects of the thumb. The median nerve-branches and the tendons involved in the scar were considered beyond repair. The ring finger was therefore sacrificed: its flexor superficialis was fashioned into a thumb flexor and its flexor profundus into an index flexor. Its glabrous skin was transferred, with a neurovascular pedicle containing its ulnar innervation and vascular supply, to the volar surface of the index. When examined seventeen years later, the thumb, index and little finger showed, by their texture and colour, that they were being extensively used. The middle finger, whose long flexors and extensors and interossei were intact, but which lacked innervated glabrous skin, was being used 'hardly at all'. It made no contribution to the grip of the whole hand. The patient said 'it does not follow'.

In reconstructive surgery of the hand, all possible cutaneous innervation is preserved. 'Particularly valuable is the experience from the work performed for surgical salvage of those hands in which only "bits and pieces" are left, but which can nevertheless be brought to perform many useful functions in daily life provided that sufficient sensory skin cover remains.'[12] Among the 'many useful functions' is that of emotional contact. 'The smallest grip with sensation is always better than the best prosthesis' was one of the aphorisms of Sterling Bunnell.

Another of his aphorisms was 'If you have nothing, a little is a lot.' 'A lot' is supplied by a 'residual hand' with just one 'finger' able to grip against its side, both surfaces being covered with sentient skin, providing the sort of 'key grip' used by normal people when unlocking a door with a key held between the pulp of the thumb and the side of the index. If even this much cannot be recoved from the 'bits and pieces', Moberg says that 'a much reduced, but still very useful function is created by splitting the forearm into two "fingers",' both covered with sentient skin. Even in the dark this Krukenberg arm 'can work with great speed and skill.' The threshold of two-point discrimination has been observed to improve from 16 mm, the normal value for forearm skin, to 8 to 10 mm.

This is not the place for any extensive discussion of the problems of powered artificial hands. Bioengineers have designed pneumatically- and electrically-powered artificial arms of marvellous sophistication, controlled mechanically or myoelectrically by surviving proximal muscles. The problem of feedback across the machine–man interface, however, remains formidable. Moberg feels that hand surgery as a specialty has not contributed as much as it might have done to the development of hand prostheses. In his experience an artificial hand providing only a key grip, powered as well as controlled by the two rays of a Krukenberg arm, 'is of course much inferior in gripping function to the Krukenberg arm itself but is still better than any other system of hand prosthesis known today.' One patient, however, whose other arm had been amputated at shoulder level, would use it only for cosmetic and social reasons, not when he was alone at home. 'I can do ten times more with the Krukenberg grip than with the prosthesis.'[13] The status of this prosthesis is that of pliers or forceps, which do not have sentient tips.

'Trials have already shown that it is possible to transfer signals of passive motion and position to skin receptors from a movable part of the prosthesis'. Such a crossing of the machine–man interface makes great sense physiologically. Moberg's opinion is that 'this will be the method in prosthesis-work in the future: to raise afferent impulses to the level of conscious perception.' For without conscious perception, the learning of new skills and their transfer to the 'more automatic' status that is 'the hallmark of manual skill and dexterity' will be impossible.[12]

CHAPTER IX

Cortex and hand

It is time to confront the brain. The lecture promised an outline of experiments on the contribution of receptors in muscles, tendons, skin and joints to the automatic governing and accurate perception of the voluntary movements of our fingers. That outline has now been sketched. Sherrington's idea 'that not only the cortex, but the whole sensory path from the periphery to the cortex cerebri, is in action during voluntary movement' has been corroborated by those experiments, and by further observations and experiments on the disordered movements of men and monkeys when the receptors have been disconnected from the brain.

A second introductory promise has still to be redeemed: an outline of experiments on the cortical neuronal hardware that receives and processes information coming in from the hand, and on the corticospinal neurones that most directly control the elemental movements of the fingers. Increasingly versatile usage of the hands in the series of living primates has depended on the evolution of this hardware, not on improvements in musculoskeletal structures that have remained morphologically primitive.

1. *Brain and hand*

The consummate refinement of tactile discrimination at the 'somaesthetic macula' has gone along with the evolutionary enlargement and differentiation of the related areas of the neocortex. So has the enormously-enriched repertory of movements of the fingers, and their motor-sensory combination in Active Touch. The areas to be considered in this chapter are all relatively closely linked to the receptor organs and skeletomotoneurones. *Enlargement* by itself implies no more than a greater bulk of hardware: *differentiation* implies an increase in *organization*. On the afferent side it implies selective, analytical extraction of 'key-features'—features important in particular contexts of perception and performance—from the welter of raw information streaming in from the receptors of skin and muscles. On the motor side it implies flexible combination and recombination of the contractions of many muscles—as prime movers, synergists and fixators—in the service of 'less automatic' performances. These fundamental analytic and synthetic functions, carried on in neuronal hardware not very far removed from the periphery, are the very foundation of our highest cognitive and perceptual-motor skills. Those skills are the product of learning extending over months and years; and of activities in other neocortical areas, subcortical forebrain structures and cerebellum, in touch with the 'special senses' as well as with the somaesthetic, but more remote from the receptors and effectors than the areas to be considered in this chapter.

Another lecture would be needed to correct the lopsidedness of a story which will largely ignore the basal ganglia, cerebellum and thalamus, in spite of their probable importance in the programming of voluntary movement.[1] Throughout the lecture brain programmes were, indeed, repeatedly invoked, but disappointingly little is known of their neuronal hardware or of its whereabouts in the brain.

It is well known that K. S. Lashley's long 'search for the engram' went unrewarded.[2] In particular he did not find it in the precentral motor cortex, the area to which this chapter will be giving pride of place. He trained monkeys to open three types of puzzle-boxes to get food, and then ablated the precentral gyri bilaterally. The initial paralysis was followed by considerable recovery of movements of the arms and hands. When the boxes were again presented it was found that the monkeys had retained their learning and, in spite of their clumsiness, went correctly about the task of opening the boxes and getting out the food.

By recording from an array of electrodes applied to the scalp, widespread electrical changes have been detected in the human brain in advance of an elemental voluntary movement, flexion of the index finger.[3] The changes related to a single rapid flexion of the finger would be lost in the background electrical activity, but if the records of several hundred self-paced flexions are averaged by a computer, the localization of the electrical events leading up to and continuing during the movement can be followed. The picture is one of initially widespread, bilateral activity, homing down on to the region of the central sulcus of the hemisphere contralateral to the finger. A slowly increasing negative wave ('readiness potential'), distributed widely and bilaterally over the brain, begins about 850 ms before the start of the movement. 'Pre-motion positivity', also widely spread, starts about 85 ms before the electromyogram of movement. A rapidly increasing additional negativity ('motor potential'), with its maximum amplitude over the arm area of the hemisphere contralateral to the finger, starts about 60 ms before the start of the electromyogram and continues during the movement. Presumably this whole sequence is the electrical sign of the simplest possible ballistic programme in action.

Searching the brains of monkeys with microelectrodes for neurones firing impulses in advance of trained movements has found them in cerebellar cortex, cerebellar nuclei, basal ganglia, thalamus and cerebral cortex. But of the possible organization of these neurones into functional assemblages which we could recognize as motor programmes we have no knowledge.

2. *Cortical 'receiving', 'association' and 'projection' areas: macroscopic aspect*

Programmed movement (Fig. 3), wherever and however it may be programmed in the forebrain and cerebellum, depends on current information (or a fresh memory trace) of the starting positions of the parts of the body and of the successive stages that bring the movement to its goal. This information is received and processed in several distinct areas of the parietal cortex (Fig. 20). A parasagittal slice through the arm areas (Fig. 20a, AB) shows that most of their total area lies buried in the walls of the central, intraparietal and superior temporal sulci (CS, IPS and STS, Fig. 20b). The areas are basically similar in men and monkeys but are much larger in men. The relatively very large proportion of the postcentral areas given over to the monkey's hand was discovered when potential waves evoked by touching the hand and fingers were first recorded from exposed postcentral cortex over forty years ago. In his Sherrington Lectures on *The Cerebral Control of Movement*, Denny-Brown described the effects of experimental lesions as 'a severe and permanent defect in all delicate spatial adjustments of movements of the hand . . . The small eversions, abductions, rotations that enable precise palpation and exploration or withdrawal depend on the integrity of pre- and postcentral gyrus.'[4]

The somaesthetic function of man's parietal cortex was established by electrical stimulation nearly seventy years ago.[5] The stimulation aroused paraesthesiae which were referred to specific peripheral loci. The large areas of reference to hand and fingers were mapped in greatest detail by Penfield in the nineteen-thirties. The effects of postcentral lesions depend on their site and size, and are well known

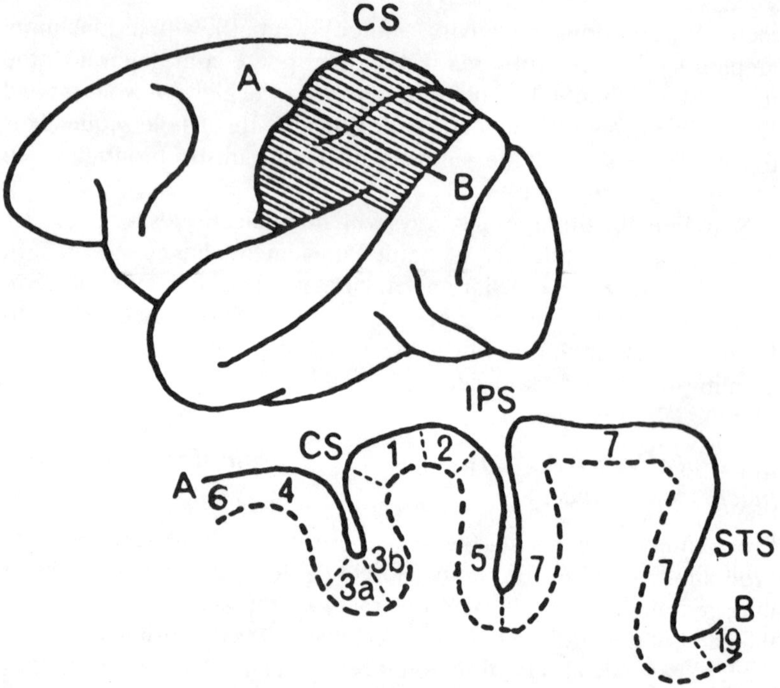

Fig. 20 Cortical areas related to the hand. The location of the parasagittal slice AB is marked on the outline of the left hemisphere of a monkey's brain. Numerals are Brodmann's. Areas 3a, 3b, 1 and 2 together constitute the primary somaesthetic receiving area, S I, the cortical terminus of the lemniscothalamo-cortical projection. (SII, not shown, is buried lateral to Area 7. It is the cortical terminus of the spinothalamic projection and concerned more with passive than active touch). CS, central sulcus; IPS, intraparietal sulcus; STS, superior temporal sulcus. See text. (Gordon, *Active Touch*, 1978, Oxford: Pergamon Press)

to clinical neurologists.[6] Impairment of sense of position and passive movement, two-point discrimination, tactile localization, reading of numbers or letters traced on the skin, and discrimination of objects by active touch, may be limited to the opposite hand. The extreme case is total neglect of the opposite half of the body: people thus affected fail to dress themselves on the affected side, to put on make-up or to shave that side of the face. In bed they may see the affected limbs, believe that they belong to someone else and demand that that other person be removed. Their drawings of people, buildings and maps are incomplete on the corresponding side.

In front of the central sulcus lie areas 4 and 6 (Fig. 20, a,b). Electrical stimulation of area 4 finds a large hand area in monkeys and men. As Sherrington ascended the primate scale with his experiments, he found that smaller and smaller 'local items of movement' could be evoked from larger and larger areas by near-threshold stimuli—an advance 'more evident in baboon than in macaque, in gibbon than in baboon, and in chimpanzee, orang and gorilla than in gibbon.' The most differentiated 'local items' were movements of individual fingers.

Penfield's Sherrington Lectures on *The Excitable Cortex in Conscious Man*[7] surveyed the large sub-areas from which differentiated movements of the opposite hand, and especially of individual fingers, could be evoked. These conscious patients knew that they had not *willed* these responses to stimulation. They were due to blunderbuss-like discharges of impulses excited by artificial electrical stimulation of the top end of the corticospinal pathway. The action of the stimuli on the cortex itself was *paralyzing*. 'The effect of the electrode is to interfere with the patient's ability to make voluntary employment of the cortex near the electrode. He may be able to move the foot and face at will, but he cannot direct the movement of the hand while hand cortex is being stimulated. Sometimes the electrode produces no movement and then this interference is the only effect of the stimulating current.' The subtly-patterned neuronal activities of selection and combination of outputs are evidently disrupted by such electrical interference—just as post-central stimulation maps *paraesthesiae*, not natural sensations.

When thinking about the increasingly large areas of cortex devoted to the fingers as one ascends the primate scale, it is amusing to recall an idea of Hughlings Jackson's, now over a century old. 'The parts whose muscles are represented in the greater number of *different* movements would be represented by the greater *quantity* of grey matter.'[8]

The corticospinal pathway from area 4 is very greatly enlarged in man. Answering a letter from F. M. R. Walshe in 1947, Sherrington wrote: 'I think the Py. (pyramidal tract) is such a *human* feature. In cat and dog it is a poor thing, in rabbit it is not easily traceable at all below the middle of the spinal brachial enlargement. In the common small monkey there is more, but not *much*. Rothmann's cutting of the pyramids years ago in the dog caused little lasting defect. In macaque the degree of recovery is surprising, when contrasted with the degree and persistance of the human condition, which is often permanent, is it not?'[9] Severe paralysis may be unaccompanied by sensory impairment, as in 'pure motor hemiplegia'. The paralytic effects range in severity from clumsiness of independent movements of the fingers of the opposite hand to loss of all voluntary movements of the opposite half of the body. Surprising as the degree of recovery in the 'common small monkey' may seem, it is never complete. There remains a permanent inability to winkle morsels of food out of holes in a board with the index finger and to take hold of them with the precision grip of thumb and index. If the pyramidal tracts are cut at birth this ability is never learned, as it is by normal infant monkeys during the first year of life.[10]

Area 6 lies immediately in front of area 4, and is now attracting much fine research. Its lateral part is probably (among other things) the penultimate link of a complex cortico-cortical pathway from the occipital lobe to Area 4 and involved in the visual orientation of the contralateral hand and fingers as they approach a target. Rupture of this pathway does not prevent the bringing of the hand on to the target, which is performed by proximal muscles which can be visually controlled from either hemisphere. But the correct orientation of the thumb and index finger (necessary if the monkey is to winkle the food out of the target) is normally controlled through the hemisphere contralateral to the hand, and now has to be discovered by trial-and-error palpation, that is, by Active Touch.[11]

The medial part of area 6, extending on to the medial surface of the hemisphere, is known for historical reasons as the 'supplementary motor area'. It is not 'supplementary' at all , but of great importance in the bilateral coordination of asymmetrical movements of the arms and hands. The better name 'supramotor'[12] has been proposed for it, implying that its status in the hierarchy of levels of motor control is higher than that of area 4. The effect of a unilateral ablation of this area is remarkable. If a normal monkey is confronted with a horizontal slab of perspex with holes drilled in it, each stuffed with a raisin, it holds one hand cupped below a hole and pushes the raisin downwards with the index finger of the other hand, so that it falls into the palm. Monkeys with a unilateral lesion of the SMA cannot do this, even though vision is allowed. They may put the index finger of one hand into the hole from above, that of the other into it from below, and push from opposite sides. In one year there was no recovery of the ability to carry out coordinated bimanual performances requiring a different contribution from each hand.[13]

Evidence that the SMA is active in voluntary movement has come from measurement of increases in regional cerebral blood flow. The idea that local brain metabolism fluctuates with functional activity, and that its products excite local vasodilatation, was set forth by Roy and Sherrington in 1890.[14] They had recorded the volume of the brains of anaesthetized dogs, using the skull as a plethysmograph, and recorded the systemic arterial and venous pressures simultaneously. The brain volume varied *passively* with changes in the systemic pressures. But intravenous injection of lactic acid, or of a filtered emulsion of dog's brain, caused an increase in brain volume that was *active*, in that it occurred independently of changes in the systemic pressures. 'We conclude then, that the chemical products of cerebral metabolism contained in the lymph which bathes the walls of the arterioles of the brain can cause variations in the calibre of the cerebral vessels; that in this re-action the brain possesses an intrinsic mechanism by which its vascular supply can be varied locally in correspondence with local variations of functional activity.' Their paper declared their explicit interest in functional localization.

In the new Scandinavian experiments[15] the regional blood flow through the 'supplementary' and motor areas was measured, following an inhalation of ^{133}Xe, by on-line processing of the activity of an array of 254 detectors applied to the skull, and displayed as a colour-map on a television monitor. The task set to the subjects was a rather tricky sequence of finger movements which had to be carefully memorized and then carried out throughout the 45 s of data acquisition. When the fingers contralateral to the detectors were performing the task the detectors overlying the SMA and the motor arm area registered an increase in blood flow. When the ipsilateral fingers were performing the SMA detectors, but not the motor arm area detectors, registered an increase.

In a second experiment the subjects were asked to rehearse the contralateral finger movements internally, making no actual movements. Electromyography certified that no movements were made. The SMA detectors, but not the arm area detectors, registered increased flow.

In a third experiment the subjects simply compressed a spring repeatedly with the contralateral index finger. The arm area detectors registered an increase, the SMA detectors no increase.

The researchers were modest in their claims—'no more than heuristic diagrams in which no true scientific accuracy is possible'—but their evidence strongly suggests that the SMAs are active bilaterally when the fingers of one hand are performing a 'least automatic' task, or even rehearsing it in mental imagery, but are inactive when a 'more automatic' task is being run off. It seems possible that technical advances in nuclear magnetic resonance and positron emission tomography will enable a finer resolution of local brain metabolism—temporal as well as spatial—in the not very distant future.

3. *Cortical 'receiving', 'association' and 'projection' areas: microscopic aspect*

When Hughlings Jackson was writing about 'grey matter' in the 1870s it was still a relatively amorphous substance. The Cell Theory had been going for forty years, but the full extent of the dendritic and axonal arborization of nerve cells was unknown, and the formal enunciation of the Neurone Doctrine was still a quarter of a century ahead. It was known that grey matter contained cell-bodies of different sizes, and Jackson speculated about their role in movement, normal and epileptic. 'Hitzig and Ferrier's points of excitation ... have many very large cells. I should, for theoretical reasons, expect ... that those which represent movements in small muscles (eyes, face and hands) will have cells comparatively small—smaller than the cells superintending movements of the large muscles of the limbs, but not so small as the cells of the very highest centres.'[16] So the '*comparatively small*' cells would be contained in the '*greater quantity* of grey matter' in which 'the greater number of *different* movements would be represented.' Jackson was groping towards correlations of function with cellular structure which are at the forefront of research today. But no accurate microscopical basis for such ideas came into being until the turn of the century, when the full splendour of cortical neurones was revealed by the Golgi method (see Fig. 23), and the science of cortical architectonics was born.

The areas labelled with numbers in Fig. 20 (and the other neocortical areas) were being mapped microscopically in Liverpool by A. W. Campbell[17] at the same time as Grünbaum and Sherrington were mapping electrically the 'excitable cortex' of the great apes. The areas were differentiated by microanatomical criteria, cytoarchitectonic and myeloarchitectonic. Campbell was Pathologist to the Asylums Board of the County of Lancaster. The Board supplied him with the means for research in its Laboratory at what was then called the Rainhill Asylum. In addition to his human material, Campbell had the cerebral hemispheres of two chimpanzees and one orang-utan from Sherrington's experimental series. 'It has been satisfactory to find that the delicate method of electrical excitation employed in stimulating the cortex in those researches has not in any way damaged the brains for histological purposes.' He also thanked Sherrington

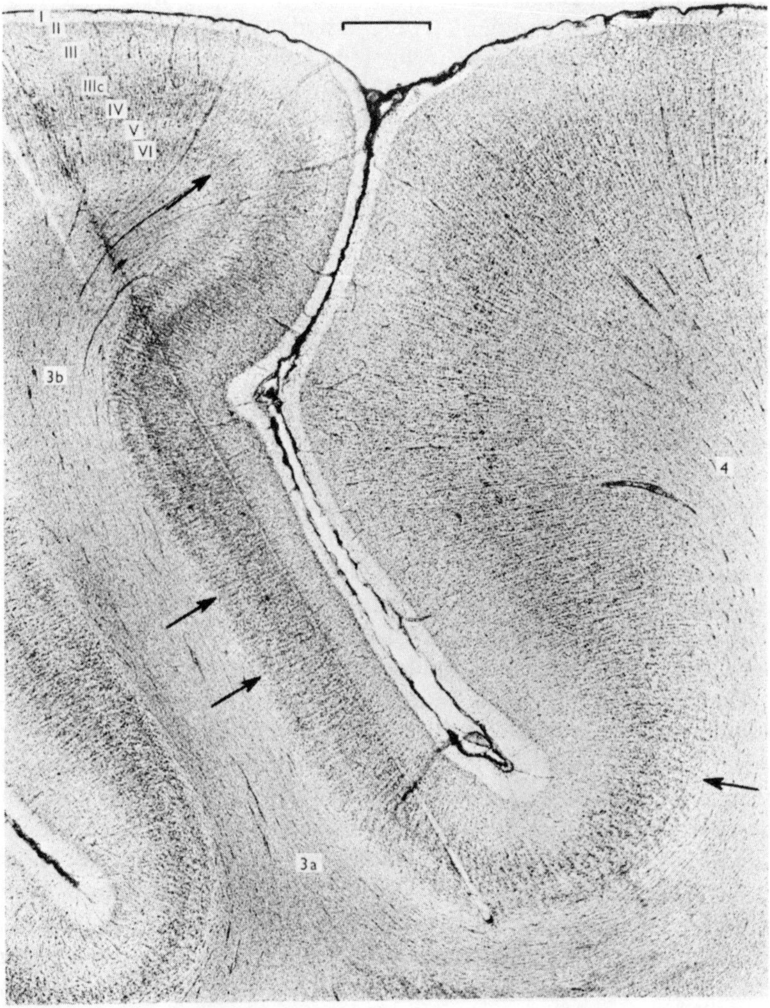

Fig. 21 Areas 3b, 3a and 4 of a baboon, stained by Nissl's method for perikarya. Horizontal bar = 1 mm. See text. (Heath *et al.*, *J. Physiol. Lond.* **257,** 199–227, 1976)

for 'counsel when in doubt, for encouragement when energy flagged and for many other kind offices.'

The major architectonic subdivisions of the cortex were differentiated quite independently by Campbell and Brodmann, whose first publications both appeared in 1903. The numbers in Fig. 20 are Brodmann's. Each area is characterized by its overall thickness, the thickness of the six laminae, the laminar distribution and packing density of the neuronal perikarya (cell bodies), and the varying bulk of the horizontally-running myelinated axons in the inner and outer bands of Baillarger. About 65% of the neurones are pyramidal neurones which send their axons out into the subcortical white matter and beyond, and the remainder are intracortical (stellate) inter-neurones of varying morphology. Campbell illustrated the great architectonic contrast between the cortex in front of and behind the central fissure. These differences are obvious in Fig. 21, a photograph of a parasagittal section through the arm areas of a baboon stained by Nissl's method to show the perikarya and the lamination.[18] In spite of the apparently greater cellularity of area 3b, a cylinder of cortex cut at right angles to the pia-covered surface of area 4 would contain about the same number of perikarya as a cylinder of the same diameter cut from area 3b.[19] Area 4 is the thickest of all the cortical areas. The very low density of perikaryon-packing means that its neurones are separated by large masses of 'neuropil'—a tangle of dendritic and axonal arborizations which would provide a very rich and flexible synaptic *connectivity* between the neurones, appropriate for the delicate control of skilled movements. In contrast, Area 3b is a relatively hard-wired topographical map of the body surface, for whose function such a degree of flexibility would seem inappropriate.

Campbell's great work is entitled *Histological Studies on the Localisation of Cerebral Function*. By itself, however, histology can discover nothing about function. It can differentiate the architectonics and map the boundaries of areas whose function is already known from clinicopathological or physiological research. But the differentiation of histologically identifiable areas whose function we do not yet know is a challenge, for if it is not axiomatic that different structures have different functions we may as well give up brain research. Increasingly subtle disorders of function, physiological and behavioural, are being detected by new and delicate tests, and should be ascribable to destruction or disconnection of specific areas. In this way, functions may become attached to hitherto 'silent' or 'association areas'. In man, modern imaging techniques can localize, during life, circumscribed lesions in 'silent areas', though it cannot of course give their cytoarchitectonic map references. Such areas will remain, initially, 'black boxes', though it will be important to have discovered their approximate whereabouts. The fundamental task is to map the details of *neuronal connectivity* within and between the boxes. This mapping is now being pushed forward by the thrust of powerful new techniques. A brief survey of the scope of modern experiments on the microarchitecture of cortical connectivity will show how the 'grey matter' is being teased apart and laid open to inspection. And in the paragraphs that follow the survey, statements about connectivity will be based on the evidence of one or more of the techniques that are about to be surveyed.

Today's mapping of neuronal connectivity is both anterograde, tracing the connections from their cells of origin towards the target neurones, and retrograde, working backwards from the target areas to the cells of origin. The classical anterograde method is that of Wallerian degeneration. Axons whose parent cells have been destroyed experimentally, or which have been cut at some point along their course, degenerate distal to the lesion which has deprived them of their gene-controlled integrity. From the eighteen-eighties to the nineteen-fifties, histology could reveal only the degenerating myelin sheaths, not the degenerating axons themselves. In the cortex this limited the scope of the method to the course of the myelinated axons; their degenerating unmyelinated arborizations and presynaptic terminals, as well as those of the far more numerous non-myelinated axons, remained invisible. The invention of the Nauta method, which stains degenerating axons (the staining of intact axons being suppressed), was the start of a new period of intensive research. The degenerating axons were made visible as far as their terminations, and at about the same time it became possible to identify their degenerating synaptic terminals under the electron microscope.

In the human cortex the mapping of neuronal connectivity by Wallerian methods requires post-mortem examination of patients with small, well-circumscribed, stationary lesions. Unfortunately the time-scales of the methods impose limitations which are not often respected by the period of survival. And even in those cases in which the requirements can be met, the sheer size of the human brain may prohibit the labour and expense of a detailed neuroanatomical investigation. For these reasons, progress depends on collateral research in monkeys, in which well-circumscribed experimental lesions can be placed in identifiable cytoarchitectonic areas.

The newest methods are not, in any case, applicable to man. They depend on the flow of axoplasm in living axons. Radioactive amino-acids injected amongst a population of neuronal perikarya are transported by cellulifugal ('anterograde') axoplasmic flow to their terminal arborizations, where they can be made visible in histological sections by autoradiography.[20] Horseradish peroxidase (HRP), injected into electrically-identified individual axons not far from their terminal arborizations by an intracellular-recording micropipette, is transported 'anterogradely' into the arborizations where it is made visible in the sections by labelling with the diaminobenzidine reaction product (Figs. 24 and 25).

Retrograde (cellulipetal) axonal transport of HRP or suitable dyes, injected into an area of axonal terminations, gives good pictures of the perikarya of the parent neurones.[21] And intracellular injection of HRP into the soma of an electrically-identified neurone, for example, an Area 4 neurone which has been identified electroanatomically by the recording of antidromic impulses sent into it from its axon in the electrically-stimulated pyramidal tract, often fills the soma, dendrites, initial segment of axon and recurrent axon collaterals apparently completely, giving a picture which loses nothing in comparison with a well-impregnated Golgi image.[22]

Another new method labels neurones whose carbohydrate metabolism has been increased, by naturally or artificially stimulated activity, for about half-an-hour before death. This method detects the increased neuronal metabolism directly; not indirectly, as in the human experiments with ^{133}Xe, by measuring changes in regional cerebral blood flow. Glucose and 2-deoxyglucose are competitive substrates for both transport and hexokinase-catalysed phosphorylation; but whereas glucose-6-phosphate is metabolised to CO_2 and water, 2-deoxyglucose-6-phosphate is not. The hyper-active neurones in which 2-[^{14}C] deoxyglucose was accumulating during that half-hour can be made visible by autoradiography of histological sections of the brain.[23]

The complexity of the human cortex is at first sight totally baffling. Every mm^2 of its surface covers about 10^5 neurones. Some of these are known to receive about 6×10^4 synapses. But it has been made very slightly less unapproachable by the discovery that it is made up of radially-orientated neuronal modules of essentially similar construction, of which man has more and mouse has fewer. These have been compared to 'the integrated circuits in electronics technology.'[24] It has long been known that the apical dendrites of the corticofugal pyramidal neurones, the incoming thalamo-cortical and cortico-cortical axons, and the chains of dendrites and axons of many of the stellate interneurones, are orientated *radially*, that is, at right-angles to the cortical laminae. The first evidence of spread of neuronal activity along radial axes came from microelectrode experiments on the somaesthetic cortex of cats and monkeys. The radial elements were at first called 'columns', but as later work in other areas has found that these are not always cylindrical, but may be wavy slabs, 'module' may be a better word. 'The functional unit is a vertical column of cells extending across the cortical layers, for each cell of such a column is related to nearly identical receptive fields, and all are excited by stimuli of the same quality, and at least in the initial stages of response these cells all discharge nearly identical patterns of impulses.'[25] Such a column would be coaxial with the thalamocortical axons supplying its input.

There was at first no microscopical evidence of the existence of *columnar* as distinct from merely *radial* structure. The search for visible inter-columnar frontiers was pursued with vigour, and soon discovered microscopically-visible structures in the striate cortex of monkeys (Area 17) and the somaesthetic cortex of mice.[26] The visible structures were limited to lamina IV in each area: to the alternating ocular-dominance columns of Area 17 in the monkey and the whisker barrels in the somaesthetic cortex of the mouse. The terminal arborizations of the thalamocortical axons are densest in lamina IV. In areas in which no such circumscribed thalamocortical columns could be made visible, an earlier proposal remains interesting: that 'the anatomical columns are not distinct, separable morphological entities. The tangential overlap is considerable . . . The physiological column results from a focus of effective stimulation of a number of definite units by the input, coupled with the inhibition of surrounding units.' And in another formulation: 'The same cortical elements could serve as a component of one functional column at one time and of another column at another time. Functional columns could form and dissolve at different times. It is obvious that the emphasis here is on the dynamic instead of the static properties.'[27]

The search for the basic *anatomical* unit of cortical architecture, the 'true cortical column' or module, has led to the discovery that it is centred on *cortico-cortical* and not on thalamocortical afferents. 'Anterograde' methods (axoplasmic transport, Wallerian degeneration) show that 'tens to perhaps several hundreds' of cortico-cortical afferents converge on each *afferent* module. 'Retrograde' methods

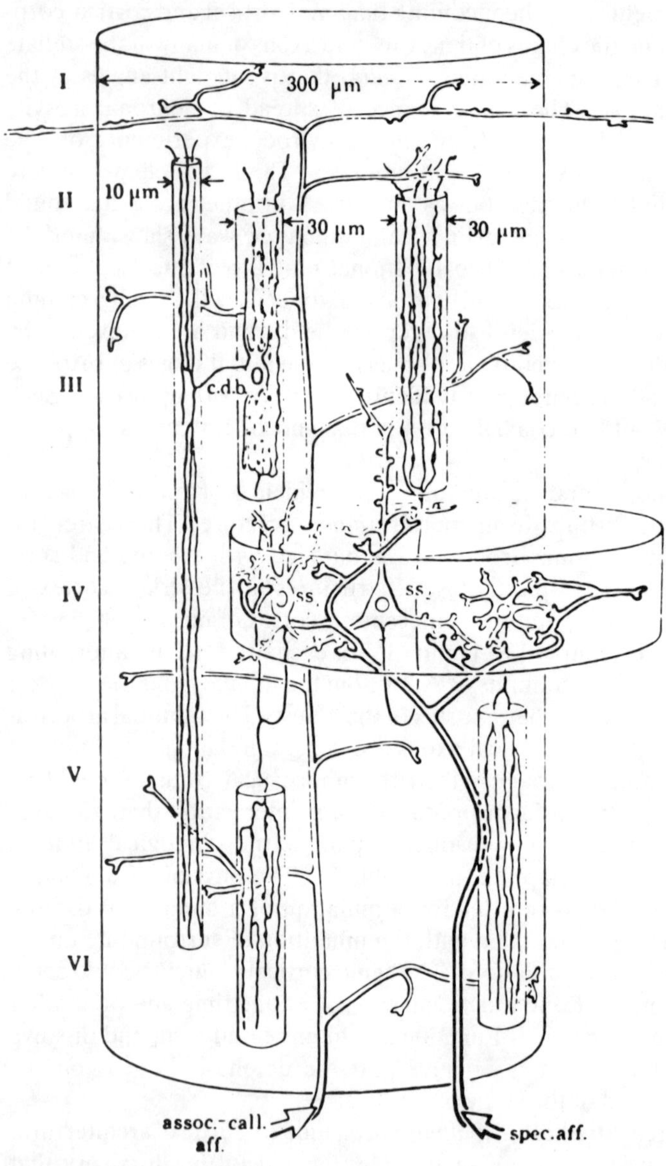

Fig. 22 Stereodiagram of a cortico-cortical module—a 300 μm diameter cylinder cut from the whole thickness of a receiving area of the neocortex. Corticofugal pyramidal neurones are omitted from the diagram, but their apical dendrites would be encased within vertical cylinders of which a few only are included in the diagram. See text. (Szentágothai, *Proc. Roy. Soc. Lond.* B. **201,** 219–48, 1978)

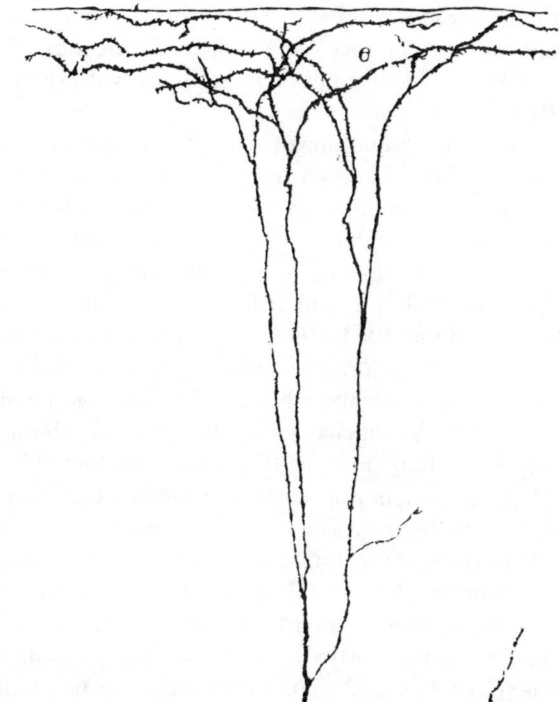

Fig. 23 Pyramidal neurone from human cortex (Ramón y Cajal). In Area 4, the perikaryon in lamina V would be 2.5 mm deep to the terminal apical dendrites in lamina I; in the much thinner cortex of mouse, the corresponding depth would be 0.7 mm (Rockel *et al.*, *Brain* **103,** 221–44, 1980)

(axoplasmic transport of HRP) show that the corticofugal pyramidal neurones which give rise to cortico-cortical axons are themselves collected together into *efferent* modules, with their perikarya in lamina III.[28]

Fig. 22 is a 'generalized' diagram of a cortico-cortical module, which shows a cortico-cortical afferent ('association and callosal'), a thalamo-cortical afferent ('specific'), and a few stellate interneurones of varying morphology.[28] No corticofugal pyramidal neurones are shown for the sake of clarity (but can be inserted imaginatively from Fig. 23). Such a module (diameter 300 μm) would contain the perikarya of some 7×10^3 neurones. The cortical laminae are numbered on the left (see also their numbering in Fig. 21). The cortico-cortical afferents give off branches in all laminae and extend as far as lamina I, where they branch extensively, spreading about 2.5 mm rostro-caudally from their point of bifurcation, in some cases as far as 4.5 mm. Their main synaptic targets in lamina I are the distal dendrites of pyramidal neurones.

The densest terminal arborizations of the thalamo-cortical afferents are found in lamina IV, as 'laminar—sometimes almost two-dimensional—termination patterns.' It is not known whether these are co-axial with the cortico-cortical modules, whether they are eccentric (as shown in Fig. 22), or whether there is no regular relation between their axes. Injection of HRP into the axons of antidromically identified thalamocortical neurones in cats[29] finds that they branch to form two or three separate arborizations in lamina IV of the somaesthetic cortex. One of the arborizations may be as far as 5 mm distant from another. Such a neurone may transmit signals from one very small receptive field on the tip of a cat's forepaw. It will be interesting to know whether the small receptive fields on the tips of a monkey's fingers are projected to more than one cortical column, or whether their distribution will turn out to be more restricted than in the cat.

Within lamina IV the thalamocortical arborizations excite, monosynaptically, the dendrites of pyramidal and stellate neurones. The latter's axons can be assumed to distribute excitation radially across the whole thickness of the cortex, because microelectrode explorations find (as we have seen) that the responding neurones are not confined to lamina IV. In Area 17 the 2-deoxyglucose method finds ocular dominance neurones active in all laminae. The morphology of the stellate interneurones suggests that their axons create, with the pyramidal dendrites they enfold, *sub-modules* whose diameters range from 10 to 30 μm (Fig. 22). These diameters correspond to those encountered in experiments with microelectrodes. Thus in Area 17 the well-known orientation columns are 30 to 50 μm in diameter, the ocular dominance 'columns' are slabs 400 μm wide. The whisker barrels of the mouse are 100 to 400 μm in diameter. The axons of stellate interneurones ascend to lamina I, in which their horizontal ramifications are extensive.

In addition to the *radial* connectivity there is also *tangential* connectivity—*parallel* to the cortical laminae. Over and above the 'mutual interpenetration' between adjacent thalamocortical modules (horizontally-spreading pyramidal dendrites and recurrent axon collaterals in laminae II to VI, and the wide ramifications of thalamocortical axons, corticocortical axons and pyramidal dendrites in the enigmatic lamina I), there are stellate interneurones with horizontally-running axons. These form synaptic baskets around the perikarya of pyramidal neurones in laminae III and V. Their axons run for 2 to 3 mm, mainly parasagittally, in lamina IV. Their synaptic endings are of putative inhibitory type, with symmetric contacts and flattened vesicles. But electroanatomical experiments show that excitation as well as inhibition is propagated tangentially in the cortex.[30] Of the parent cells of the horizontally-running myelinated axons in the bands of Baillarger rather little is known. Their myelin protects them from impregnation by the Golgi method. Some at least are intrinsic in origin, since they do not degenerate after undercutting a slab of cortex.

In building up this scheme of the cerebral cortex in general, intuitive piecing-together of Golgi pictures has played an important part. It has thrown up ideas which can now be tested in modern anterograde, retrograde, metabolic and immunocytochemical experiments. Such experiments hold great promise for the next few years.

We are accustomed to speak of 'receiving', 'association' and 'projection' areas. But the modules of *all* areas—and not only those of Flechsig's classical 'association areas'—receive converging corticocortical and thalamocortical inputs, 'integrate' (or 'associate') them, and are centred on corticofugal pyramidal neurones which project their integrated outputs to cortical or subcortical targets. In the firing of pyramidal neurones, therefore, we might find a key to the operation of the module. There is ample reason to believe that modules are engaged in disseminating and redistributing information. But beyond this, it has been suggested that they may function as 'detectives', investigating, so to say, 'suspicious coincidences' between co-varying environmental events.[31] Some coincidences would signify imminent possibilities of reward, others imminent danger of dissolution. The firing of pyramidal neurones might signal the detection of a significant coincidence between two or more inputs to a module. Their sensitivity to anything 'suspicious' in a particular coincidence would be expected to vary in accordance with the prevailing environmental situation and the animal's 'expectation' or 'set'. The responsiveness of a module to a particular coincidence might be determined by selective setting of the excitability of the apical dendrites of its pyramidal neurones in lamina I—possibly by monoaminergic 'state control' by the locus caeruleus and other widely-diffused subcortical projections to that lamina.

4. *Inputs from the hand*

A brief microelectrode tour of the areas of Figs. 20 and 21 will give some idea of the kinds of things that are being discovered about the cortical processing of afferent information from the monkey's hand, and will raise questions about the underlying neuronal connectivity on which it depends.

Let the tour begin in Area 3a. Until the nineteen-sixties there was no evidence that signals from intramuscular receptors reached the cortex; it was therefore argued that the muscle spindles could not be responsible for any sensations of position and movement, and that these must therefore be due to signals from the joints. Electroanatomical experiments on the arm areas of baboons then showed that the arm region of area 3a does receive signals from muscle spindles, including those of the forearm and hand.[32] Fig. 21 shows the track of a microelectrode reaching down to this narrow and deeply-buried part of the arm areas. The track is easy to see because the glass micropipette was left *in situ* during fixation and embedding. Different neurones responded to selective electrical stimulation of muscle afferents in the ulnar, median and deep radial nerves; some 'deep radial' neurones were also excited by rapid stretching or longitudinal vibration of the tendons of an isolated muscle (extensor digitorum communis). The afferent pathway is in the dorsal columns. A small proportion of the neurones also responded to cutaneous afferent volleys. The finger region of Area 3a is immediately adjacent to that of Area 4, and sends cortico-cortical axons into it. These could contribute to the servo-control of movement. Signals arousing kinaesthetic perceptions, truthful or illusory, would be projected from Area 3a to the parietal cortex.

Architectonically, area 3b rather superficially resembles the primary visual area (Area 17) in its relative thinness, well-marked lamination and 'cellularity'. Both are primary receiving areas. Neurones in Area 3b respond to mechanical stimulation of small receptive fields, particularly on the glabrous skin of the fingers.[33] The fields resemble those of single axons mapped by microneurography (VIII). They overlap one another. Some neurones nearest to area 3a, but definitely in area 3b, respond to muscle as well as to skin. It is noteworthy that area 3b contains rather few cortico-cortical neurones that can be labelled by retrograde transport of HRP injected into Area 4. So the relatively 'raw' input from small cutaneous receptive fields is projected caudally for further processing, rather than rostrally towards the motor output. The main route from Area 3b to Area 4 would pass through Area 5, the 'somaesthetic association area', which has been shown, by retrograde transport to HRP, to project to Area 4. Antidromically-labelled corticospinal neurones of area 4 do receive a rich input from the skin of the fingers, but most of this probably comes through a thalamocortical, not a cortico-cortical route.

The topographical map is not a simple hard-wired structure. It is dynamically maintained. If an afferent nerve is cut, or a finger amputated, an underlying map of other areas of skin is immediately unmasked in the deprived territory. This is gradually followed by some rearrangement in the parts of the map at the edges of the territory. Subsequent reinnervation, by the same nerve or by a transposed nerve, leads to a gradual reinterpolation of the reinnervated part.[34] The old palimpsest is restored by a new one.

Topographical location of a tactile stimulus to the hand is not only a datum in its own right, it is also the first stage in a complex sequence of processing. Particular 'linking features' of the stimulus-pattern are extracted from the signals from many overlapping receptive fields and projected to other areas in which the *features*, not primarily the *topography*, are separately represented. Such arrangements are called serial or hierarchical processing. To the extent that some of the primary input is projected to more than one area we can also speak of parallel processing.

The microelectrode explorations of Area 3a, and many of those of 3b, were done on monkeys under light general anaesthesia. To follow the next stages of processing we require the cooperation of trained monkeys working for reward. At a preliminary operation under general anaesthesia a miniature microelectrode assembly is screwed to the skull. Humane observers agree that the wearing of this implant for the rest of the monkey's life is painless and without significant discomfort. At the end of the experiment the brain is required for relating the tracks to the cortical laminae and the cyto-architectonic areas. The microelectrode explorations are spread, in short sessions, over many weeks, so that very large amounts of data can be collected from each monkey.

As the exploring microelectrode is moved from Area 3 towards Area 5, it encounters neurones whose impulse-discharges are related less to the topographical map and more to the non-topographical representation of features extracted from, and common to, stimuli applied to wider areas of skin.[35] Tested by merely probing the skin, and with nothing beyond topographical localization in mind, such neurones would be supposed to be responsive to much larger receptive fields than those picked up further rostrally—just a coarser-grained topographical map. But their real significance becomes apparent when it is found, for example, that they respond to the movement of a point across the large receptive field in one direction but not in another. Yet other neurones respond to movement of an edge but not of a point. Such responses represent, non-topographically, information extracted from more than one topographically-receptive field crossed by the point or edge. Other neurones discharged impulses when the monkey took hold of a solid rectangle but not a cylinder, and yet others when the cylinder, but not the rectangle, was grasped. Such higher-order neurones may represent information extracted from proprioceptive inputs (Area 3a), from exteroceptive inputs (Area 3b), and from motor outputs, in a context of 'active touch' (Chapter VIII).

There are neurones which respond by a discharge of impulses when a joint is moved in one direction but not in the other, whether the movement is active or passive. These are 'kinaesthetically-truthful neurones'. From human experiments on the perception of finger movements (V and VI) we should expect that this feature has little to do with inputs from joints; rather that it has been *extracted* from the inputs from skin and muscle. The question whether such neurones would continue 'truthful' under conditions in which people experience erroneous kinaesthetic *illusions* has already been raised (VII 2, 3; Fig. 11).

Area 5, the 'somaesthetic association area', contains neurones which fire impulses in relation to still more complex combinations, for example when the hand is brought to the mouth, or when the hand is rubbing the foot in a particular direction. In Area 7, there are neurones which discharge when the monkey's gaze is directed towards an interesting object or when he reaches for it with his hand.[36] Some at least of the neuronal hardware for the coordinated movements of hand and gaze in peripheral and macular vision must be here, and must be in touch with the motor apparatus of Areas 4 and 6 by pathways not yet worked out.

Area 6 lies in front of Area 4. Neurones in its lateral part are active in movements towards a target, or attention to a target, in either half of visual space. Their activity seems to be related more to the organization of the movement itself than to its visual guidance.[37] Those in the medial part of Area 6 (the supplementary motor area so-called) are active in relation to movements of either arm, and receive no very direct inputs from the periphery.

The microelectrode tour of the areas has now reached its terminus in Area 4. *En route* the microelectrode interrogated a sample of the neurones in the various cytoarchitectonic areas. They could be categorized by their responsiveness to particular features, topographical or non-topographical, of a simple or 'compound' stimulus. Some responded to *passive* stimuli to the skin or joints of the hand, others were *active* in tactile and stereognostic exploration. Each neurone would have been located at some point on the afferent or efferent side of a module which would itself have formed part of complex intra- and interareal networks. Its perikarya would have been located in one of the cortical laminae, its axon and dendrites orientated radially but also with some horizontal connectivity in one or more laminae. As yet we do not know enough to assign neurones of different response-types to their proper places in the modules. We do know of cases in which no correlation between a neurone's response-type and its morphology has been found.

5. *Area 4*

Though the 'giant cells of Betz' (Fig. 23) gave to Area 4 its original name 'area giganto-pyramidalis', they form but a small proportion of its total population of corticofugal neurones, which includes smaller corticospinal neurones with thinner axons and 'parapyramidal' neurones projecting to other areas of cortex and to thalamus, striatum and brain stem. All these neurones are the target for, and the 'final common path' leading out from, two great sets of neuronal networks of the forebrain and cerebellum. Of these, one set originates in wide areas of the cortex and converges on Area 4 by way of the striopallidum and ventrolateral thalamus. The other set arises from equally wide areas of cortex and converges on Area 4 by way of pontine nuclei, inferior olive, cerebellum, red nucleus and ventrolateral thalamus. In its totality this 'final common path' is the executant of control by the forebrain of most of the 'less automatic' and all of the 'least automatic' postures and movements of the body. It is helpful to think of it as receiving motor 'commands' from higher-level programmes and re-writing them in terms of appropriate combinations and sequences of muscular contractions and relaxations. But it also receives signals from the moving parts which detect, and tend to correct, deviations from the programmed trajectories and velocities of movements (Chapter IV). Area 4 has been felicitously described as 'a primary device for fast adjustment of programmed motor patterns to afferent signals.'[38] And Bernstein's 'regulator' (6, Fig. 2) seems to invite attachment to the neuronal hardware of Area 4.

Sherrington's stimulations of the 'excitable cortex' elicited coordinated movements that were remarkably similar to those he could elicit from the chronically-isolated spinal segments by stimulating areas of skin or the trunks of afferent nerves. Evidently, corticospinal and parapyramidal pathways (cortico-rubrospinal, cortico-reticulospinal) can address the complex interneuronal networks of the segmental and intersegmental reflex arcs. As he ascended the primate scale he found that smaller and smaller 'local items of movement' could be evoked by near-threshold faradic stimulation. The most circumscribed 'local items' were movements of individual fingers in the great apes. These discoveries led him to the conception of the 'excitable cortex' as an *analytic organ*, capable of breaking up by selective inhibition 'compounds already constructed by lower centres'; but also as a *synthetic organ* for motor acts, elaborating, from the coordinately-executed 'local items', 'larger combinations varied in character and serviceable for purposes of different and varied kind.'[39]

From this formidable ensemble we are now going to abstract one part only of the neuronal hardware that controls the movements of the hand: those corticospinal neurones which form direct synaptic connexions with the skeletomotoneurones that drive the muscles of the forearm and hand. This monosynaptic corticomotoneuronal (CM) pathway is a taxonomic feature of primates. It also contributes to the control of feet and (in Spider and Woolly Monkeys) of prehensile tails, whose ends are covered on the flexor aspect with glabrous skin; it bulks larger as one ascends the primate series and reaches its greatest development in man.[40] This abstraction is feasible because contemporary microanatomical and electroanatomical methods are biased in favour of large neurones with thick fast-conducting axons. Over and above its particular relevance to motor control it has value for cortical studies in general. For the corticospinal neurones may be taken as a paradigm of corticofugal neurones as 'final common paths' leading out of their cortical modules. They can be readily identified antidromically by stimulating the medullary pyramid or the lateral corticospinal tract. From their discharges, and especially from intracellularly-recorded synaptic potentials, excitatory and inhibitory, there is much to be learned about the connectivity of the cortical afferents and interneurones whose impulses converge on the output neurones of the module.

It is remarkable that the CM neurones are separated by one synapse only from the neuromuscular junctions. If they are to address the skeletomotoneurones directly, without creating chaos by their by-passing of the phylogenetically-ancient segmental apparatus of local reflexes and of locomotor and other programmes, they must, so to say, be fully informed about prevailing segmental and suprasegmental activities; and their 'commands', however powerful and pre-emptive, must not be crude, but highly structured and fully integrated with all other 'commands' descending from the forebrain. It is otherwise in focal (Jacksonian) epilepsy. The pathologically-synchronized bursts of cortical activity discharge CM volleys which by-pass the less direct pathways, causing convulsive jerks which start in those parts which have the densest monosynaptic connectivity and are therefore preferentially accessible to the cortex: thumb and index, side of cheek or big toe.

In experiments on monkeys working for reward, the output neurones can be identified by antidromic stimulation of the medullary pyramid by chronically implanted electrodes.[41] The term 'pyramidal tract neurone' (PTN) is reserved for those so identified. Not all of these would be CM neurones. Methods for *their* electroanatomical identification will be explained below. PTN fire bursts of impulses in advance of movement, and frame-by-frame analysis of movie films corroborates naked-eye observations in showing that the discharges of a particular PTN are *related* to movement in a *particular* direction at a *particular* joint or joints of the forearm and hand. Pooling the results from a collection of PTN showed that they fell into two groups, forming two 'ramps of recruitment': the first started 500 ms in advance of the movement, the number of neurones entering into activity reaching a peak 200 ms in advance of the movement, and falling to zero 200 ms after the start of movement; the second started 100 ms in advance of the movement, reached its peak 100 ms after the movement and decayed to zero after a further 400 ms.[42] Activity so far in advance of movement might merely form part of a general 'preparedness-to-move', of which the human 'readiness potential', and a body-wide facilitation of skeletomotoneurones even when the movement is going to be confined to the hand,[43] are also indices.

It would be surprising, however, if PTN were not themselves capable of actually triggering a ballistic movement, rather than merely priming the skeletomotoneurones in readiness for another trigger or triggers. Transmission across CM synapses is potentiated by repetition at high frequencies, and the discharge of bursts of impulses at high frequency is a characteristic property of PTN.[44] Pairs or brief bursts of CM impulses at higher-than-average frequencies would bring the skeletomotoneurones quickly nearer to their firing threshold. Such shorter-than-average intervals are often recorded in the output of PTN. So the priming action of steady 'ramps of recruitment' could prepare the skeletomotoneurones for triggering by pairs or bursts of impulses, synchronized within the bursts of a number of PTN converging on the same skeletomotoneurones. Each such burst would trigger electromyographic activity in hand muscles within about 11 ms of its dispatch from Area 4 in a baboon and 17 ms in a man.

If the PTN do contribute to the triggering of movements, then the output of populations of them must be focused on the skeletomotoneurones that are to be triggered. The brain's ability to learn so to focus them is remarkable. In man it is a simple matter to record the impulses fired by single motor units of hand and forearm muscles from intramuscular electrodes, and to present them to the subject in the form of a visual (cathode ray oscillograph) display. It is found that subjects can learn to select and fire single motor units, that is, single skeletomotoneurones, at will. They can only do this from amongst the skeletomotoneurones that are recruited in the gentlest voluntary contractions. The experimenters said to them: 'Now I will ask you to discharge only this tall unit', or 'Now stop the activity of all motor units except this one.' Any unit which the subject could not learn to control in half an hour was classified as uncontrollable voluntarily. Of 273 units from thenar muscles and extensor digitorum communis (whose motoneurones in baboons receive the largest totals of monosynaptic potentials from their CM colonies), 198 were controllable.[45] Such a remarkable ability to select single skeletomotoneurones argues for a wonderfully subtle routing of activity in the thick synaptic neuropil of Area 4 to pick up, in appropriate functional groupings, the required CM neurones, scattered and intermingled, as we shall see, with unwanted ones which may be just not selected or may be actively suppressed by inhibition. Something like this, rather than anything like the push-buttons disparaged by Bernstein, seems nearer to reality.

Learning to select single skeletomotoneurones from a visual display can be observed in man; learning to select single PTN must be observed in monkeys.[46] They were given an instrumental display of the electromyogram of biceps brachii and also of the discharges of a flexor-related PTN. The monkey was rewarded for increasing the firing of the flexor-related PTN and at the same time reducing the contraction of the biceps. It succeeded in this task. Whether we assume that the firing of other flexor-related neurones was being voluntarily inhibited in proportion to the reduction of electromyographic activity, or whether we assume that these other neurones were simply not being selected, we may be glimpsing something of the remarkable potentialities of the intracortical output-selecting machinery and of the neural hardware upstream that drives it.

It looks as if *different* populations of PTN can engage the *same* muscles in different performances. Many PTN remain silent while a monkey is squeezing a bulb held in the palm and fingers (power grip), but show intense activity when picking up a small seed between the tips of thumb and index (precision grip).[47] Some muscles cannot fail to have been engaged in both performances; in the power grip they must have been driven by other PTN, whether monosynaptically or by a polysynaptic corticospinal connection.

6. *On transcortical loops*

So much for the possibilities of selection by forebrain programmes of the PTN/CM outputs from Area 4. Now for the inputs to PTN from the muscles and skin of the moving hand which would provide for 'fast adjustment of programmed motor patterns to afferent signals.'

The suggestion of an automatic transcortical control loop[48] stemmed from considerations of two kinds. First, what could be the biological advantage of the very fast and direct path which monosynaptically connects the CM neurones to the skeletomotoneurones of hand muscles? The transmission times, about 11 ms in baboon and 17 ms in man, are off the bottom end of the time scale of movement at any joint. These times would however make sense if the CM neurones formed the efferent limb of a suprasegmental control loop (Fig. 11) whose afferent limb, originating in muscle receptors, could reach Area 4 by a direct lemnisco-thalamo-cortical route; by an indirect lemnisco-thalamo-cortico-cortical route *via* area 3a (Figs. 20 and 21); or *via* the cerebellum. In its automatic governing of programmed force and speed such a loop could operate in the manner already described (Chapter IV) for the phylogenetically-ancient segmental control loop (Figs. 7 and 11). The directness and power of the CM connection suggested that in the primate series there has been evolutionary advantage in incorporating its CM neurones into an automatic control network, analogous to the one attached to the skeletomotoneurones on which it has been overlaid. Secondly, a classical record of a PTN showed that its movement-related firing increased when the movement was opposed by a weight, and declined when it was assisted by gravity.[49] This suggested that the PTN formed part of a control loop. Pioneering studies of long-loop reflexes in monkeys (see below) and men (Chapter IV) soon removed the matter from the realm of speculation and placed it on a proper experimental footing.

In man, the electromyographic response of the isometrically-contracting flexor pollicis longus to a sudden stretch includes long-loop reflex components starting 40 ms and 55 ms after the stretch (responses A and B, Fig. 9). And when a ramp flexion of the interphalangeal joint of the thumb is perturbed by halting, unloading or forcibly reversing the movement, each of the three servo-type electromyographic responses which form the 'tulip' begins at the same time in each individual (about 50 to 60 ms after the perturbation: Fig. 10)—though the response to halting may come later than the other two. The possibility that these responses are mediated transcortically must now be considered.

That the long-loop response of FPL is a suprasegmental and not, or not only, a spinal response, is shown by its dependence on the integrity of the dorsal columns. It is abolished by a plaque of multiple sclerosis at the top end of a dorsal column, and restored if the lesion is healed.[50] That it passes through the cerebral cortex is made highly probable by its abolition or attenuation in the thumbs of patients suffering from cortical motor weakness, with or without cortical sensory loss. While it is fair to say that human studies have not *proved* that the long-loop response actually passes *through* the cortex, as distinct from being facilitated by the cortex at some subcortical level or levels, the transcortical story would have been demolished if the responses had survived these lesions unscathed.[51]

The ideal investigation requires small, accurately localized, curable precentral lesions, in patients whose long-loop responses could be followed from pre-lesion normality to post-operative restoration of normality. One case coming near to this ideal has been reported. A superficial tumour the size of a cherry was removed from the left central sulcus, where it had been causing cortical motor and sensory loss in the right, dominant hand. Recovery of motor function was complete within one month; sensation improved more slowly, but the patient was able to return to his dental practice with only 'minimal uncertainty' in the use of his right arm, which became symptom-free by nine months. Two days after operation he was unable to produce ramp flexion of the right thumb, but could maintain a steady position against the torque lever. Stretching of FPL evoked no response, even at segmental latency, in spite of a brisk finger-jerk. A week later there was a response to stretch at 55 ms (long-loop component B) but none at 40 ms (component A). Seven weeks after operation the first 'tulip' was recorded, with servo-type responses to Stretch and Release but none to Halt. Stretching the muscle now evoked a small component A, with component B as good as that from the left hand.

A series of patients with cortical and capsular lesions, some with cortical motor and sensory loss, some with sensory loss only, others with motor loss only, also showed impairment or absence of component A, sometimes with enhancement and prolongation of the segmental response and of component B. 'Tulips', absent initially, sometimes recovered, with or without the Halt response. There was, however, some doubt about the normality of the supposedly 'unaffected' hands, making their responses suspect as controls for those of the affected hands. In the responses to stretching the isometrically-contracting muscle, the A and B components and the later 'voluntary' component were differentially vulnerable in ways not fully understood. 'In general the results are consistent with the transcortical theory of the long-latency stretch reflex for the thumb, but, in detail, they indicate that the theory will require elaboration.' The different responses may prove to be related to more than one transcortical or subcortical channel—thalamocortical, thalamo-cortico-cortical and cerebello-thalamo-cortical. 'If there is a transcortical stretch reflex loop either it is not a simple one or there are other pathways in parallel with it.'

In baboons, the muscle spindles of extensor digitorum communis, whose skeletomotoneurones receive a strong cortico-motoneuronal projection, have been stimulated by rapid stretches and by vibration. The low-threshold afferent axons have been selectively stimulated in the nerve. The input generated by each of these three methods travels through the dorsal columns and excites neurones in Areas 3a and 4. The finger regions of these two areas are immediately adjacent to one another at the bottom of the central sulcus (Figs 20 and 21). HRP injected into Area 4 is transported by retrograde axoplasmic flow into neurones in Area 3a.[52] There is also electroanatomical evidence that PTN in Area 4 receive input from muscle receptors in extensor digitorum communis, but at a latency a few milliseconds longer than their arrival in Area 3a. In the face of these connexions in subhuman primates[53] it would be surprising if there were no such cortical substratum for part at least of the long-loop servo-type responses of the human thumb.

Experiments on monkeys working for reward have paralleled the electromyographic responses in men and fitted the PTN and other cortical neurones into the picture.[54] In the electromyographic response to a sudden stretch of a monkey's biceps there was a first burst with a latency of 12 ms, appropriate for the segmental control loop (Fig. 7). A second burst, fired at a latency of 30 to 40 ms, was timed appropriately for a transcortical loop. PTN which were active in relation to elbow *flexion*, and also discharged when the elbow was passively *extended*, fired about 24 ms after the stretch. Some post central neurones fired 10 ms after the stretch, some unidentified precentral neurones fired 14 ms after the stretch. Conduction time from cord to biceps was 7 ms: total 31 ms. These timings suggest that the monkey's transcortical loop passes through Area 3 on its way to Area 4.

The transcortical reflex could be 'gated' by the voluntary 'set' of the monkey. He was instructed by a green light to push the manipulandum away from him when the perturbation occurred (thus relaxing biceps), or by a red light to pull it towards him (thus contracting biceps). The perturbation itself stretched biceps in some trials and relaxed it in others. The second (trans-cortical) electromyographic burst was 'gated in' or 'gated out' by these instructions. The prior-instructed precentral neurones showed an appropriate increase or decrease in firing about 200 ms in advance of the perturbation to which the monkey had to respond. But the early burst was also modified in some degree by the prior instruction. There is evidence that opening or closing of the transcortical loop is under the control of the cerebellum.[55]

To function in servo-type responses, the PTN whose discharges are related to active flexion of a joint must respond, not to passive *flexion* of that joint, but to passive *extension* of it, that is, to passive lengthening of its flexor muscles. These PTN (if they had access to the sensorium) would be 'kinaesthetically untruthful'. Their behaviour contrasts with that of the 'kinaesthetically-truthful' postcentral neurones which signal 'joint is extending' whether the extension is passive or active. It is, however, appropriate for an automatic servo-type government of the output of these PTN by an input from muscle spindles driven in α-γ-linkage by the cortex, analogous to the control of the skeletomotoneurones by the segmental loops (Fig. 7). In a collection of precentral neurones (not identified antidromically) which responded to passive movements of joints, 88% of responses were to movements in the direction opposite to that in which intracortical microstimulation had actively moved the joint.[56] They could therefore have functioned in the sense required by a transcortical servo-loop.

It has become clear that the transcortical control loop does not have the function of compensating for large perturbations (Chapter IV). Rather it seems that it is highly sensitive to small changes in spindle activity such as those illustrated in Fig. 8. PTN discharging in relation to active supination of the forearm, and excited 'reflexly' by passive pronation (lengthening of the supinator muscles) respond most strongly to very small imposed deviations from the steady position the monkey is rewarded for maintaining. It is in small ramp movements that the control is most in evidence.[57]

Not everyone is convinced of the existence or importance of transcortical loops or even of long-loop reflexes.[58] The survival of long-loop reflexes in the large arm and leg muscles in decapitate monkeys has been raised as a difficulty for the transcortical theory.[59] Such survival, however, cannot prove that those muscles did not also have transcortical reflexes when their connections were intact, and certainly cannot prove that the finger muscles do not have them. Since their spindles are connected to the PT/CS neurones by a demonstrated pathway, and these neurones make monosynaptic connections with the skeletomotoneurones of finger muscles, the hypothesis that the PTN act *only* by facilitating subcortical long-loop reflexes (whose existence few people would wish to dispute) seems uneconomical. Another difficulty is that muscles oscillate mechanically when suddenly stretched, and each oscillation evokes a segmental reflex which leads to repetitive e.m.g. responses ('segmentation of the electromyogram').[60] No such segmentation, however, is seen in the servo-type Halt and Release responses to perturbation of a voluntary ramp movement of the thumb. Control experiments are beginning to sort out the long-loop components from the artefacts due to oscillation.[61]

There are grounds for asking whether there is also a transcortical loop from receptors in the glabrous skin of the fingers (the 'somaesthetic macula') as well as one from the muscle spindles. The unconscious automatic tightening of grip force when an object slips between thumb and index occurs 60 to 80 ms after the start of the slip. In monkeys the PTN whose discharges are related to movements of the fingers are readily excited by stimuli to glabrous skin. Such responsiveness would surely be at least as strongly developed in man, probably much more so. In addition to providing an automatic control loop for the very rapid adjustment of grip force to slipperiness, the PTN would form part of the wider cortico-cortical and intracortical connectivity that would be needed for the delicate voluntary control of Active Touch.

7. *Structure of the cortico-motoneuronal projection*[62]

The topography of the CM outputs that could contribute to Sherrington's 'analytic' selection and 'synthetic' recombination has been mapped electroanatomically in baboons and monkeys under general anaesthesia. Proof that it is CM outputs that are actually being mapped requires intracellular recording from the target skeletomotoneurones. Focal anodal current pulses, applied to the cortical surface, directly excite the CM axons in lamina VI. Each pulse sends a well-synchronized volley of impulses down the fastest axons in the lateral corticospinal tract which, on arriving at the spinal segment, evokes in the target motoneurones an excitatory synaptic potential at monosynaptic latency. The extent of the cortical projection area which contains the CM neurones whose outputs converge on the target cell can then be mapped by exploration with the focal anode. Given this proof that the monosynaptic CM connection exists, the minimal motor responses of hand muscles that can be mapped by intracortical microstimulation[63] can be assumed to be mediated, in large part, monosynaptically. The two methods agree in showing that each skeletomotoneurone of a baboon's hand muscles receives converging monosynaptic excitation from a *colony* of CM neurones scattered across an irregularly-shaped projection area that may be as large as 20mm^2, buried in the anterior wall of the central sulcus and spreading on to the surface of the precentral gyrus (Figs. 20 and 21). The largest quantities of monosynaptic action are projected to the motoneurones of extensor digitorum communis and the small muscles of the hand. The colonies are not uniform in texture, but contain one or more 'hot spots', where the density of the Betz cells projecting to the motoneurone must be greater. The density falls away towards the edges of the projection area. The CM colonies belonging to motoneurones of different muscles overlap freely.

As far as the CM connectivity is concerned, therefore, Bernstein's criticism of 'push buttons' (Chapter II) is upheld. Pushing a button at any 'point' would involve CM neurones belonging to more than one hand muscle, as well as those of some overlapping but thinner-textured colonies belonging to the motoneurones of more proximal arm muscles.

CS neurones monosynaptically excite the Ia-inhibitory interneurones. By adjusting the level of this excitation, a central programme could inhibit or disinhibit transmission across the segmental reciprocal-inhibitory linkage, and could block it completely in situations in which co-contraction of antagonistic muscles is required. Another projection excites fusimotor neurones (Fig. 7), in some cases monosynaptically, so that α-γ-linkage is provided for. Those α-motoneurones which receive the largest quantities of CM excitation also receive the largest quantities of monosynaptic excitation from the muscle spindles, so that in α-γ-driving, the convergent CM and Ia inputs will reinforce one another's actions on the skeletomotoneurone membrane. One is reminded of a neurosurgeon's dictum, based on the effects of surgical deafferentation (Chapter III), that the prompt and correct addressing of the cortical message depends on afferent impulses.

Does every CM neurone in a colony form part of an anatomically-circumscribed module? To the extent that the neocortex as a whole is an assemblage of *cortico-cortical* modules ('true cortical columns') (Fig. 22) the answer is yes. But every 'true cortical column' receives 'tens to perhaps several hundreds' of axons from wide areas of cortex. What should we be looking for in these inputs, and how would we recognize it in microelectrode recordings? As for the CM neurones in a 'true cortical column', we can be sure, from what we know of the overlapping of the projection areas and the intermingling of the cells of the colonies, that CM neurones with different targets will be sitting cheek-by-jowl within the 300 μm column.

Whether there are any fixed *thalamocortical*-CM modules, as distinct from 'functional columns' that 'could form and dissolve at different times', is unclear. The dense coupling between thalamo-cortical arborizations and stellate interneurones in lamina IV of Area 3b has almost no visible counterpart in the lush neuropil of the 'agranular' cortex of Area 4 (Fig. 21), in which laminae III and V, which contain the perikarya of pyramidal neurones, are greatly thickened. If there are any structural modules we have, as yet, no evidence from microelectrode explorations, tangential histology or 2-deoxyglucose of their size or shape. We only know that finger-related PTN respond regularly and reliably to touching or stroking glabrous skin, to passive movement of joint or to localized prodding of muscle bellies. These are presumably the afferent signals which provide for 'fast adjustment of programmed motor patterns'. In 80% of PTN, passive input and active output were related narrowly to the same joint. In other PTN there was convergence from a wider territory on to the motor output, e.g. from the whole hand on to the index finger. The input–output coupling is secure enough to justify provisional use of the term 'module', even if we cannot specify its size and shape.[64]

Intracortical microstimulation can also be used to define the motor target of those PTN which happen to be clustered together within a 'hot spot'. The inputs to a sample of neurones within the 'spot' can then be defined by recording their impulses with the microelectrode that was used for the stimulation. But without antidromic identification we cannot know whether these impulses are those of the PTN in whose targets we are interested, or are those of stellate interneurones or corticofugal pyramidal neurones with other targets, cortical or subcortical.

The results of these experiments taken together are in broad agreement. Topographically there is no *mosaic* in the sense that all the 'modules' related to a particular joint are circumscribed within a single sub-area and isolated from the similarly-circumscribed 'modules' of all other joints. Certainly there is no experimental support for any suggestion that the hand subdivision of Area 4 might be built up of isolated cylindrical columns 300 μm in diameter, one for every muscle in the hand. Within the overlapping CM colonies very many of the 'modules' are related to finger-joints. These 'modules' receive a fast lemnisco-thalamocortical input from the glabrous skin of the fingers, and are concentrated in the depth of the anterior wall of the central sulcus; but many are scattered nearer the top of the sulcus and on the convexity of the precentral gyrus, where they are freely intermingled with the 'modules' related to elbow and shoulder.

So far we have been considering the *convergence* of colonies of CM neurones on to single skeletomotoneurones. But since PT axons are known to branch, the projection from a CM colony may be *divergent* as well as convergent. Single PTN send collateral branches to other targets along their course: striatum, thalamus, red nucleus, pontine nuclei and inferior olive (to cerebellum), reticulospinal nuclei, dorsal column nuclei and dorsal and intermediate regions of the spinal grey matter. Not every PT axon sends collateral arborizations to all these targets. But collectively there is abundant connectivity which could provide for the gating of 'kinaesthetically-untruthful' inputs and for arousing the 'sense of effort' (Chapter VII). The branching has been displayed electroanatomically, by recording from PTN and backfiring antidromic impulses into them from one or more stations suspected of receiving corticospinal collaterals.[65] A second method depends on retrograde axonal transport of two fluorescent dyes, one injected into each of two putative target areas. One dye stains nuclei and the other cytoplasm. Staining of nucleus and perikaryon proves that the double-labelled PTN contributed axon collaterals to both targets.[66]

All CM neurones are PTN but not all PTN are CM neurones. In monkeys working for reward, antidromically identified PTN can be probed by a method known as spike-triggered averaging (STA), to find out if they are connected monosynaptically to the motoneurones of one or more muscles.[67] The action potentials ('spikes') of a PTN whose discharges were related to dorsiflexion of the wrist were recorded by a microelectrode in the arm area of the cortex. Six extensor muscles of wrist and fingers were wired for electromyography. When the monkey dorsiflexed his wrist and kept up a steady force against a load the PTN and all the muscles became active. Many other PTN would also have been firing impulses in relation to the wrist movement, and would presumably have been contributing to the driving of one or more of the muscles. The problem was to discover whether this particular PTN was contributing any monosynaptic drive to one or more of them. So every action-potential it discharged was made to trigger the recording apparatus, and the records of the cumulating averaged electromyograms were searched for any deviation from the general level of random activity. After the PTN had fired 4,000 impulses, the averaged electromyograms of three of the six muscles showed an unmistakable increase of activity, starting 5 ms after the PTN's impulses had left the cortex. This is the sum of the times taken (1) by a fast PTN impulse to travel to the spinal segment, (2) by a monosynaptic potential in a skeletomotoneurone to reach its peak, (3) by a motor nerve impulse to travel from the ventral

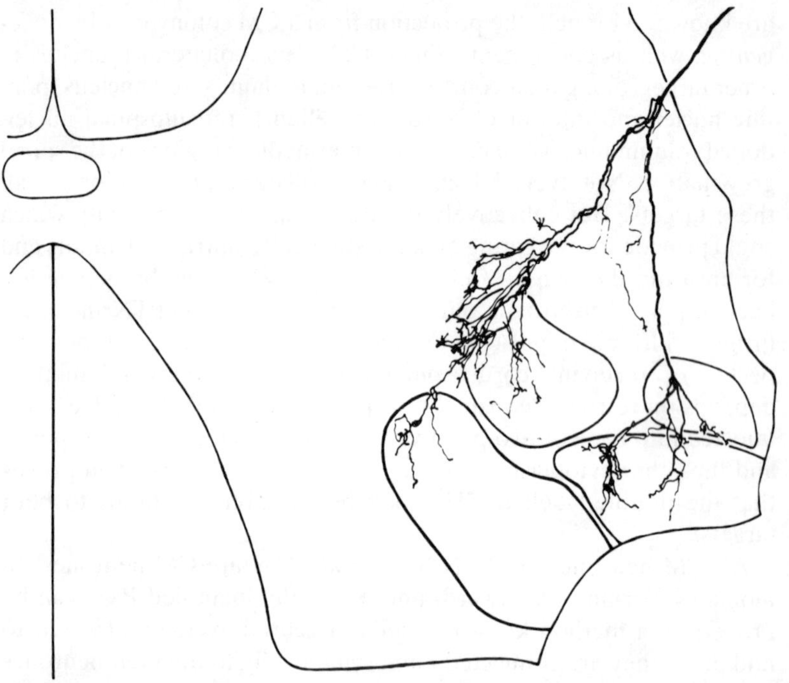

Fig. 24 Transverse ramification of cortico-spinal axon in seventh cervical segment. Upper two motor nuclei contain skeletomotoneurones of ulnar nerve; lower two, skeletomotoneurones of radial nerve. See text. (Shinoda *et al.*, *Neuroscience Letters* **23,** 7–12 (Fig. 2), 1981)

root to the neuromuscular junctions, (4) by the delay at the neuromuscular junctions, and (5) by the muscle impulses to reach the e.m.g. electrodes. So this CM neurone had monosynaptic connections with at least one skeletomotoneurone belonging to each of these three muscles. If it had any connections with a motoneurone or motoneurones in the other three muscles, the synaptic potentials it evoked in them were too small to produce any detectable post-spike-facilitation (PSF). Connections might however have become apparent if the averaging process had continued for longer: another CM neurone had to discharge 22,000 impulses to produce detectable monosynaptic PSF in two out of the six muscles. Similar experiments on finger muscles have found that CM impulses generally influence the motoneurones of one muscle only.[68]

Corticospinal axons also form profuse arborizations in more than one spinal segment (Figs. 24 and 25). This has been proved by an experiment which beautifully illustrates the power of a combination

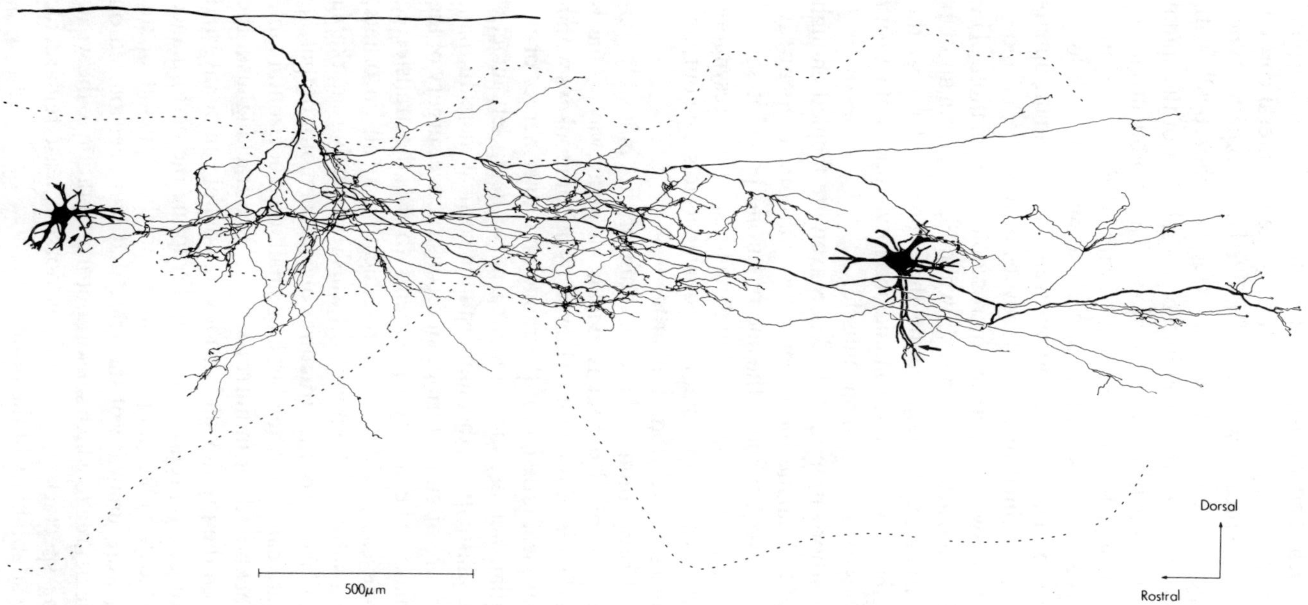

Fig. 25 Rostro-caudal arborisation of a main corticospinal collateral at C_8 level, one of several such collaterals given off by corticospinal axons running in the dorso-lateral funiculus. Two skeletomotoneurones of the ulnar nerve are shown; the outermost limits of their dendritic trees are marked by dotted lines. A thick arrow shows site of one light-microscopically identifiable synaptic contact upon each motoneurone. (Lawrence *et al.*, *J. Comp. Neurol.* **232,** 499–510, 1985)

of electroanatomical and microscopical techniques.[69] Corticospinal volleys were sent down from the hand area while the lateral corticospinal tract was being probed with an HRP-filled micropipette. When a CS axon was penetrated it was microinjected ionophoretically with HRP. The arborization of the axon in the ventral horn of the spinal cord was reconstructed from serial sections. The motoneurones of different muscle groups had also been filled with HRP by gross injection of their nerves (Fig. 24: ulnar, upper two nuclei; radial, lower two nuclei). Some of the synaptic boutons were found to make apparent contacts with proximal or distal dendrites of the labelled motoneurones, but this awaits electron microscopic confirmation. The extensive rostrocaudal arborization is shown in Fig. 25, from another experiment. Further experiments along these lines will reveal the extent of branching of CM axons to make monosynaptic contact with the motoneurones of more than one muscle.

The terminal arborization of a *single* CS axon are tangled enough (Figs. 24 and 25). Imagine the superimposed arborizations of say, a hundred CS axons, spanning the motoneurone pools of several muscles. What routing of impulses, what dynamic states of presynaptic arborizations and post-synaptic neurones enable us to single out, and fire at will, a particular skeletomotoneurone?

That the motor programmes of the forebrain are not inflexibly connected to its corticospinal output is evident in many ways. One is the remarkable degree of function achieved by infant monkeys within six months of surgical crossing of the radial and ulnar nerves or the radial and median nerves, the remaining nerve being left intact.[70] This recalls the plasticity of the somaesthetic cortical map. In man, tendon transfers are often performed in reconstructive surgery of the forearm and hand. The ability of patients to use the transferred muscles, in some cases without any re-education at all, in others, however, after a substantial amount,[71] furnishes equally remarkable evidence of the plasticity of the structures of cortical motor output.

One's signature can be written large or small, on a horizontal table or a vertical blackboard, or in mirror-writing. In a tracking task, the coupling between the joystick and its moving pointer can be inverted while the tracking is in progress without disrupting the performance of a skilled subject. Some subjects, intent on their pursuit of the moving target, were unaware of the trick that had been played on them.[72] And the degree to which sensorimotor adaptation to the wearing of inverting spectacles is possible is not the least remarkable evidence of the adaptability of the brain's output.[73]

CHAPTER X

Prospect

Enough experimental and clinical research has been brought together to illustrate the power and promise of complementary work on the uses and control of the hands of men and monkeys, each with their balancing advantages and limitations, each raising questions for the other, both contributing to a common conceptual framework.

The most refined studies of connectivity, topographical mapping of the body surface, non-topographical mapping of trigger features extracted from the welter of information available from the receptors in skin and muscle, and the structural organization of motor output from the forebrain, are possible only in monkeys. These studies will continue to suggest observations that can be made on man and to help in their interpretation, as they have been doing since the days of the pioneers, Ferrier, Horsley and Sherrington.

Protracted microelectrode experiments on the brain of man, though painless, are ruled out because of possible injury and shortening of life. But for combinations of microneurography, electroencephalography, gross and single-fibre electromyography, goniometry and neuropsychological research on skilled movement man has the advantage that, if unambiguously instructed, he can perform immediately, or can learn to perform, tasks which monkeys can perform, if at all, only after weeks of patient training by reward. Most important will be an expansion of work on patients with circumscribed lesions of cortical subcortical or segmental structures. The lesions can now be located with fair accuracy by modern imaging techniques as well as by clinical and neurophysiological examination, but should be completed, whenever possible, by microanatomical investigation.

The gains will be by no means limited to basic science. Basic science is important in its own right as part of man's cultural heritage, but also because it so often leads, often after many years, to important and unforseen practical applications. Understanding how the brain controls the movements of our hands, our perceptions of them, and the faculty of active touch is bound to be helpful in planning the habilitation of people damaged at birth and in the rehabilitation of injured adults. The plasticity of somaesthetic and motor maps will have to be understood before it can be controlled for the benefit of patients.

In other cases the central nervous system is intact but eyes or hands or ears are damaged. In the planning of reconstructive surgery, the devising of motor prostheses, and the invention of sensory substitutions, the adaptability of the brain must be exploited to the full. Let no-one underestimate its capacity to use whatever sources of information or strategies may be available to it in different behavioural contexts or 'sets'— or feel frustrated when the result of some apparently straightforward experiment is equivocal because the brain has learned to circumvent the conditions imposed by the experimenter.

Notes

CHAPTER 1

1 C. S. Sherrington (1899). On the spinal animal, being the Marshall Hall Prize Address. *Transactions of the Royal Medical and Chirurgical Society*, **82,** 449–77. London: H. K. Lewis.

2 C. Bell (1833). The Hand: Its Mechanism and Vital Endowments as Evincing Design. *The Bridgewater Treatises, IV*. London: William Pickering. See pp. 245–65, A comparison of the eye with the hand.

Sherrington's Presidential Address 'The Relation between Structure and Function, as Examined in the Arm', read to the Liverpool Biological Society on 11 November 1898 (*Transactions of the Liverpool Biological Society*, **13,** 1–20), does less than justice to Bell's Treatise. 'Its stately language and wealth of illustration will long preserve it as a masterpiece of popular exposition of science. Yet every Chapter seems far from us as a bygone age; its pages are alive no longer. How has this come about? The facts are true, and therefore living as ever; the style is classical, and therefore never dying. It is the point of view that has suffered change. The question asked throughout is the question that it was the fashion at that time for biologists, and especially anatomists, to ask and to attempt to answer. . . . We are content to struggle with a humbler problem. The question *why?* is not answered by positive science, but only the question *how?* and sometimes the question *how much?* . . . To confuse natural science with metaphysics is a mischievous mistake for the enquirer, at least for the present and for some centuries to come.'

It is true that Bell's book comes to us across the great gulf created by *The Origin of Species*. But it does not read like a Tract, and a modern biologist may well feel that Bell paid little more than lip service to the 'as evincing design' of his Title. The task had been assigned to him by the President of the Royal Society of London, with the advice of the Archbishop of Canterbury and the Bishop of London, under the Will of the Rt Hon and Reverend Francis Henry, Earl of Bridgewater. Bell was one of eight persons who were each 'appointed to write, print and publish one thousand copies of a work 'On the Power, Wisdom and Goodness of God, as manifested in the Creation'. The Will specifically mentioned 'the construction of the hand of man.' Bell says in his Preface that when he undertook this task, 'he thought only of the pleasure of pursuing these investigations, and perhaps too little of what the public were entitled to expect from an Essay composed in circumstances so peculiar, and forming a part in "this great argument".'

The twenty pages of 'comparison of the eye with the hand' deserve to be more widely known as one of the great classics of the *how*, not the *why* of neurology. They, and the other passages I shall quote, are innocent of propaganda.

3 A. L. Yarbus (1965), cited by A. R. Luria (1973). *The Working Brain*. London: Allen Lane, The Penguin Press. See Figs. 31 and 66.

4 G. Holmes (1936): Looking and Seeing; (1938): The Cerebral Integration of the Ocular Movements. In *Selected Papers of Gordon Holmes* (ed. C. G. Phillips). Oxford: Oxford University Press, 1979, 468–79.

5 C. D. Marsden, P. A. Merton and H. B. Morton (1981). Human Postural Responses. *Brain*, **104,** 513–34.

6 N. Bernstein (1967), *The Co-ordination and Regulation of Movements*. Oxford: Pergamon Press.

7 R. Granit (1973). Demand and Accomplishment in Voluntary Movement. In *Control of Posture and Locomotion* (eds. R. B. Stein, K. B. Pearson, R. S. Smith and J. B. Redford). New York: Plenum Press.

8 J. R. Napier (1956). The Prehensile Movements of the Human Hand. *J. Bone. Jt. Surg.*, **38B,** 902–13.

9 A. H. Schultz (1961). Some Factors influencing the Social Life of Primates in general and of Early Man in particular. In *Social Life of Early Man* (ed. S. L. Washburn). Chicago: Aldine, 58-90.

10 J. C. Eccles (1975). Letters from C. S. Sherrington F. R. S. to Angelo Ruffini between 1896 and 1903. *Notes and Records of the Royal Society of London.*, **30,** 69–88.

11 F. W. Mott and C. S, Sherrington (1895). Experiments upon the Influence of Sensory Nerves upon Movement and Nutrition of the Limbs. *Proc. Roy. Soc. Lond.*, **57,** 481–88.

12 C. G. Phillips, S. Zeki and H. B. Barlow (1984). Localisation of Function in the Cerebral Cortex: Past, Present and Future. *Brain*, **107,** 327–61.

13 C. S. Sherrington (1885). On Secondary and Tertiary Degenerations in the Spinal Cord of the Dog, *J. Physiol. Lond.*, **6,** 177–91.

14 A. S. F. Leyton and C. S. Sherrington (1917). Observations on the Excitable Cortex of the Chimpanzee, Orang-utan and Gorilla, *Q. Jl. exp. Physiol.*, **11,** 135–222.

CHAPTER II

1 Jackson's leading ideas are summarized by C. G. Phillips and R. Porter (1977) in *Corticospinal Neurones, Their Role in Movement.* London: Academic Press. 3, 9, 322-27.

2 In Sherrington's Foreword to the 1947 Edition of *The Integrative Action of the Nervous System*, xiv-xv. Cambridge University Press.

3 R. S. Creed, D. Denny-Brown, J. C. Eccles, E. G. T. Liddell and

C. S. Sherrington (1932). *Reflex Activity of the Spinal Cord:* Oxford: Clarendon Press.

4 Bernstein, op. cit (I, 6). For a century, since the pioneering work of Marey, Muybridge and Braune & Fischer, refined methods of recording human and animal movements have been available, but the possibility of correlating such records with neuronal activities is less than half a century old. The gap narrowed when simultaneous records of locomotion and electromyograms of sampled muscles were made from intact, unconstrained animals (A. Lundberg (1969). Reflex control of stepping. *The Nansen Memorial Lecture*, The Norwegian Academy of Science and Letters. Oslo: Universitetsforlaget) and men (e.g. E. Knutsson and C. Richards (1979) Different Types of Disturbed Motor Control in Gait of Hemiparetic Patients. *Brain*, **102,** 405–30. Long lengths of flex are nowadays giving way to telemetry.

5 A. Brodal (1973). Self-observations and neuroanatomical considerations after a stroke. *Brain*, **96,** 675–94.

6 Foerster, O. (1931). The cerebral cortex in man. *Lancet*, **2,** 309–12.

7 J. E. Desmedt and E. Godaux (1978). Ballistic Skilled Movements: Load Compensation and Patterning of the Motor Commands. In *Cerebral Motor Control in Man: Long Loop Mechanisms*. Prog. Clin. Neurophysiol., **4,** ed. J. E. Desmedt, Basel: Karger, 21–55.

CHAPTER III

1 The neuronal mechanisms of the cat's predatory reaching and grasping behaviour have been greatly elucidated in the last few years, and are now rather well understood: B. Alstermark, A. Lundberg, U. Norrsell and E. Sybirska (1981). Integration in descending motor pathways controlling the forelimb in the cat. 9. Differential behavioural defects after spinal cord lesions interrupting defined pathways from higher centres to motoneurones., *Exp. Brain Res.*, **42,** 299–318.

2 M. Jeannerod (1981). Le contrôle de l'oeil sur le geste. *La Recherche*, **12,** 376–378; Intersegmental coordination during reaching at natural visual objects. In *Attention and Performance* (ed. J. Long and A. Baddeley) Hillsdale: Erlbaum. 153–68.

3 Mott and Sherrington (see Note I.II).

4 R. E. Coggeshall, M. L. Appelbaum, M. Fazen, T. B. Stubbs and M. T. Sykes (1975). Unmyelinated axons in human ventral roots, a possible explanation for the failure of dorsal rhizotomy to relieve pain. *Brain*, **98,** 157–66. They cite O. Foerster's earlier challenge to the Bell-Magendie law.

5 C. S. Sherrington (1931). Quantitative Management of contraction in lowest-level co-ordination. *Brain*, **54,** 1–28.

6 J. Bossom (1974). Movement without proprioception. *Brain Research*, **71,** 285–96.

7 O. Foerster (1927). Schlaffe und Spastische Lähmung. In *Handbuch*

der normalen und pathologischen Physiologie (eds. A. Bethe, G. V. Bergmann, G. Embden and A. Ellinger), **10,** 900–01. Berlin: Springer.

8 W. R. Gowers and J. Taylor (1899). *A Manual of Diseases of the Nervous System*, Vol. 1, 3rd edition, 450. 'The great fact of the pathology of tabes is that it is a neural degeneration in the sensory nervous system, peripheral and central.' (p. 474).

9 J. C. Rothwell, M. M. Traub, B. L. Day, J. A. Obeso, P. K. Thomas and C. D. Marsden (1982). Manual Motor Performance in a Deafferented Man. *Brain*, **105,** 515–542.

10 M. Jeannerod, F. Michel and C. Prablanc (1984). The control of hand movements in a case of hemianaesthesia following a parietal lesion. *Brain*, **107,** 899–220.

CHAPTER IV

1 It is easy to forget, when concentrating on the antagonistic arrangement of FPL and EPL at the interphalangeal joint of the thumb, that the arrangement is not as simple as it looks at first sight. For FPL can also flex the metacarpophalangeal joint, and if the task is to flex the interphalangeal joint, the metacarpophalangeal joint must be kept stationary by isometric co-contraction of extensor pollicis brevis and some of the thenar muscles, acting as 'synergists' to FPL as 'prime mover'. More proximal muscles of forearm, arm and trunk act as 'essential fixators' and 'postural fixators' (C. E. Beevor (1904). *The Croonian Lectures on Muscular Movements and their Representation in the Central Nervous System.* London: Adlard). The greater the force exerted by FPL, the greater the spread of activity to the more proximal muscles.

2 Rothwell *et al.*, loc. cit. (III, 9).

3 J. E. Desmedt and E. Godaux (*loc. cit.*, II, 7). See their pages 40–43. It seems possible that in a brisk ballistic movement, transmission across the relay nuclei of the afferent pathway is inhibited by a well synchronized 'corollary discharge' in corticospinal axon collaterals (cf. **IX,** 57 and pp. 63, 121).

4 Fig. 5 shows that the patient's approach to the target differed from that of control subjects. The force generated by the patient reached the target without overshoot in about 0.5 s. The force generated by the control subject overshot the target level, then fell back to it, also in about 0.5 s. These differently-shaped approaches invite comment. To the control subject, response feedback was available from the force-measuring Golgi tendon organs. The slight initial excess of force generated by the programme (Fig. 5) could have been trimmed down to the target level in response to this feedback within the time available. The patient's approach could have been controlled by internal feedback from corticospinal collaterals, progressively increasing in intensity until it had signalled, without overshoot, that the force demanded by the short-term trace of the visually-trained programme had been reached.

5 For full information see P. B. C. Matthews (1972). *Mammalian Muscle Receptors and their Central Actions*. London: Arnold.

6 Because of the elasticity of long tendons, this relation cannot be constant for all muscles under all conditions of tension, length and rate-of-change of length. See P. M. H. Rack and H. F. Ross (1984). The tendon of flexor pollicis longus: its effects on the muscular control of force and position at the human thumb. *J. Physiol., Lond.*, **351,** 99–110.

7 K.-E. Hagbarth and Å. B. Vallbo (1969). Single unit recordings from muscle nerves in human subjects. *Acta physiol. scand.*, **76,** 321–34.

8 R. Granit (1955): *Receptors and Sensory Perception*. New Haven: Yale University Press, 268; (1968): The functional role of the muscle spindle's primary end organs. *Proc. R. Soc. Med.*, **61,** 69–78; (1975): The Functional Role of the Muscle Spindles—Facts and Hypotheses. *Brain*, **98,** 531–56.

9 Å. B. Vallbo (1970): Slowly-adapting muscle receptors in man. *Acta physiol. scand.* **78,** 315–33; (1973): Muscle spindle afferent discharge from resting and contracting muscles in normal human subjects. In *New Developments in Electromyography and Clinical Neurophysiology* (ed. J. E. Desmedt), **3,** Basel: Karger, 251–62.

10 Vallbo comments on the absence, in this experiment, of 'a systematic change which was closely related to the muscle length. Thus it seems that the sense organ was relatively insensitive to muscle length. An alternative interpretation is that there was a continuous increase in the fusimotor outflow which compensated for the muscle shortening.'

11 The small oscillations of Fig. 8C (upper record: spindle frequency) are inversely related to the small oscillations in velocity (lower record: time derivative of the joint angle signal). Vallbo comments: 'The two signals go in opposite directions in most instances, implying that the faster the muscle was shortening, the lower was the discharge rate and *vice versa* . . . the excitatory drive on to the skeletomotorneurones decreases immediately when the movement progresses faster and increases when the movement progresses slower. In this way any variations in the speed of movement—which could be due to a variation of the load, a variation in the frictional resistance or an irregular skeletomotor output—would be reduced in amplitude and the result would be a smoother movement.' Such automatic stabilizing of velocity could be mediated by segmental and suprasegmental reflexes.

12 T. Hongo, A. Lundberg, C. G. Phillips and R. F. Thompson (1984). The pattern of monosynaptic Ia-connections to hindlimb motor nuclei in the baboon: a comparison with the cat. *Proc. Roy. Soc. Lond.* B., **221,** 261–89.

13 P. B. C. Matthews (1972), *op. cit.* (IV, 5), Chapter 10. The idea that the operation of α-γ-linkage extends to reciprocal inhibition is due to T. Hongo, E. Jankowska and A. Lundberg (1969). The Rubrospinal Tract II. Facilitation of interneuronal transmission in reflex paths to motoneurones. *Exp. Brain. Res.*, **7,** 365–91.

14 J. F. M. Clough, D. Kernell and C. G. Phillips (1968). The distribution of monosynaptic excitation from the pyramidal tract and from primary spindle

afferents to motoneurones of the baboon's hand and forearm. *J. Physiol. Lond.*, **198,** 145–66.

15 G. Holmes (1946). *Introduction to Clinical Neurology*. Edinburgh: Livingstone, 98.

16 C. D. Marsden, P. A. Merton, H. B. Morton, J. E. R. Adam and M. Hallett (1978). Automatic and Voluntary Responses to Muscle Stretch in Man. In *Cerebral Motor Control in Man: Long Loop Mechanisms. Prog. Clin. Neurophysiol.*, **4** (ed. J. E. Desmedt), 167–77. Basel: Karger. This volume is a fertile source-book of facts and of the differing interpretations that promote progress and prevent fossilization of ideas. See J. C. Houk, *Participation of Reflex Mechanisms and Reaction-Time Processes in the Compensatory Adjustments to Mechanical Disturbances*, 193–215.

17 C. D. Marsden, P. A. Merton and H. B. Morton (1984). New observations on the human stretch reflex. *J. Physiol. Lond.*, **360,** 51P.

18 Experiments on a variety of responses at 'long-loop' latency are revealing the degree of their dependence on prior instruction. A subject is maintaining a steady flexor force against a lever with the terminal phalanx of the thumb. The usual long-loop response to a sudden radialward tug on the wrist is a silent period in the e.m.g. of FPL, due to the unloading of its spindles. But if the subject has been instructed to maintain contact with the lever the silence is replaced by a burst of motor output. This overriding response is due to the action of the pull on the *shoulder*. So if your wrist is jerked in the radialward direction at a party, this prior-instructed long-loop 'grab reflex' should save you from dropping your glass. Its latency is too short to be voluntary (M. M. Traub, J. C. Rothwell and C. D. Marsden (1980). A Grab Reflex in the Human Hand. *Brain*, **103,** 869–84). Marsden *et al.*, *op. cit.* (I, 5) give further examples of 'the ability of the system we are dealing with to evaluate complex situations and to adapt the rapid automatic responses of the subject in an apparently purposive way.' See also the monograph *Neurophysiological Approaches to Higher Brain Functions* by E. V. Evarts, Y. Shinoda and S.P. Wise. New York: Wiley, 1984, for much stimulating discussion of perceptual-motor 'sets'.

19 C. D. Marsden, P. A. Merton, H. B. Morton, J. C. Rothwell and M. M. Traub (1981). Reliability and efficacy of the long-latency stretch reflex in the human thumb. *J. Physiol. Lond.*, **316,** 47–60.

20 C. D. Marsden, P. A. Merton and H. B. Morton (1976). Servo action in the human thumb. *J. Physiol. Lond.*, **257,** 1–44.

CHAPTER V

1 Bell on the Law of the Roots (*op. cit.*, I, 1, 193–4): 'It was this conviction—that we are sensible of the action of the muscles, which led me to the investigation of their nerves . . . two classes of nerves,—that on exciting one of these, the muscle contracted, that on exciting the other, no action

took place. The nerve which had no power was found to be a nerve of sensation: and thus, it was proved that there is a nervous circle connecting the muscles with the brain. ... It was only by examining the nerves at their roots ... that I succeeded in demonstrating their distinct functions.'

2 The word 'kinaesthesis' was coined by H. Charlton Bastian in 1880 for use in a somewhat different sense, in the context of his idiosyncratic theory of the initiation of movement. He supposed that a voluntary movement is preceded by a central ('ideal') revival of its 'kinaesthetic' image, and that it is guided by 'kinaesthetic' impressions that influence the motor discharge, although in a subconscious way. He regarded the cerebral cortex as efferent but not motor. In his view the only motor structures were those of the brainstem and spinal cord. He thought that 'the full blaze of consciousness' is aroused 'only in new or unfamiliar voluntary action.' His final statement is in *Brain*, **32,** 328–41 (1909).

3 D. O. Hebb (1966). *A Textbook of Psychology*, 2nd ed. Philadelphia: Saunders, 252. See also M. Critchley (1953). *The Parietal Lobes*, Chapter VIII: Disorders of the Body-Image. London: Arnold; and R. Melzack and P. R. Bromage (1973). Experimental phantom limbs, cited by P. B. C. Matthews (1982, 210: *op. cit.*, V, 9).

4 C. S. Sherrington (1900). The Muscular Sense. In *Textbook of Physiology* (ed. E. A. Schäfer), Vol. 2, 1002–1025. Edinburgh: Young J. Pentland.

5 C. G. Phillips *et al.*, *loc. cit.* (I, 12).

6 R. Granit (1972). Constant Errors in the Execution and Appreciation of Movement. *Brain*, **95,** 649–60. A. G. Feldman and M. L. Latash (1982). Afferent and Efferent Components of Joint Position Sense: Interpretation of Kinaesthetic Illusion. *Biol. Cybern.*, **42,** 205–14.

7 C. S. Sherrington (1900), *loc. cit.* (V, 4)

8 K.-E. Hagbarth (1983). Microelectrode Exploration of Human Nerves: Physiological and Clinical Implications. *J. Roy Soc. Med.*, **76,** 7–15.

9 M. Hulliger, E. Nordh, A.-E. Thelin and Å. B. Vallbo (1979), cited by P. B. C. Matthews (1982): Where does Sherrington's 'Muscular Sense' originate? Muscles, joints, corollary discharges? *Ann. Rev. Neurosci.*, **5,** 189–218. See pp. 194, 196.

10 M. Knibestöl (1975). Stimulus-Response Functions of slowly-adapting mechanoreceptors in the human glabrous skin area. *J. Physiol. Lond.*, **245,** 63–80. See his Fig. 4B, p. 70.

11 A. Brodal, *loc. cit.* (II, 5)

12 W. R. Gowers and J. Taylor, *op. cit.*, 18 (III, 8)

CHAPTER VI

1 M. J. Cross and D. I. McCloskey (1973). Position sense following surgical removal of joints in man. *Brain Res.*, **55,** 443–45.

2 Again, it is easy to forget, when thinking about experimental pulling

of tendons in relation to kinaesthesis at finger joints, that the long muscles of the fingers act across more than one joint. For example, the tendons of flex. dig. profundus cross the radiocarpal, carpometacarpal, metacarpophalangeal, proximal and distal interphalangeal joints. As John Hunter (cited by C. E. Beevor, *op cit.*, IV, 1) pointed out in 1777, unwanted actions at one or more of these joints must be opposed by one or more muscles (Beevor's 'synergists'). Signalling by intramuscular receptors of movements at one only of these joints must be correspondingly complex.

3 G. M. Goodwin, D. I. McCloskey and P. B. C. Matthews (1972). The Contribution of Muscle Afferents to Kinaesthesia shown by Vibration Induced Illusions of Movement and by the Effects of Paralysing Joint Afferents. *Brain*, **95,** 705–48.

4 D. I. McCloskey (1978). Kinaesthetic Sensibility. *Physiol. Rev.*, **58,** 763–820. See pp. 783–85.

5 E. Moberg (1983). The Role of Cutaneous Afferents in Position Sense, Kinaesthesia and Motor Function of the Hand. *Brain*, **106,** 1–19.

6 D. I. McCloskey, M. J. Cross, R. Honner and E. K. Potter (1983). Sensory Effects of Pulling or Vibrating Exposed Tendons in Man. *Brain*, **106,** 21–37.

7 E. Moberg, *loc. cit.* (VI, 5).

8 P. B. C. Matthews (1982). Where does Sherrington's 'Muscular Sense' originate? Muscles, Joints, Corollary Discharges? *Ann. Rev. Neuroscience*, **5,** 189–218. See p. 198.

9 P. B. C. Matthews (1972), *op cit.* (IV, 5), Chapters 4 and 5.

10 Goodwin *et al.*, *loc. cit.* (VI, 3).

11 Matthews (1982), *loc. cit.* (VI, 8), page 199.

12 P. B. C. Matthews (1977), cited by D. I. McCloskey (1978, *loc. cit.*, VI, 4, 786).

CHAPTER VII

1 Most of our knowledge of the actions of CS axon collaterals on somaesthetic 'relay' nuclei comes from experiments on the cat; some of the seminal papers are reviewed by Phillips and Porter (*op cit.*, II, 1), pp. 111–17.

2 G. M. Goodwin *et al.* (1972), *loc. cit.* (VI, 3).

3 The spindles of finger extensor muscles which are actively maintaining a particular joint angle are *stataesthetically-uninformative*. There was no significant difference in their firing frequency whether the joint was being accurately held in either of two positions, 20° apart (M. Hulliger, E. Nordh and Å. B. Vallbo [1982]. The absence of position response in spindle afferent units from human finger muscles during accurate position holding. *J. Physiol. Lond.*, **322,** 167–79). The spindles were evidently pre-empted in the service

of automatic control. During voluntary relaxation, however, they *were* responsive to passively-imposed changes in joint angle.

4 G. Eklund and K. E. Hagbarth (1966). Normal Variability of Tonic Vibration Reflexes in Man. *Expl. Neurol.*, **16,** 80–92.

5 W. Z. Rymer and A. D'Almeida (1980). Joint Position Sense: the Effects of Muscle Contraction. *Brain*, **103,** 1–22; P. E. Roland and H. Ladegaard-Pedersen (1977). A Quantitative Analysis of Sensations of Tension and of Kinaesthesia in Man: Evidence for a Peripherally-Originating Muscular Sense and for a Sense of Effort. *Brain*, **100,** 671–92.

6 D. N. Rushton, J. C. Rothwell and M. D. Craggs (1981). Gating of Somatosensory Evoked Potentials during Different Kinds of Movement in Man. *Brain*, **104,** 465–91.

7 R. Granit (1972), *loc. cit.* (V, 6).

8 Goodwin *et al.* (1972), *loc. cit.* (VI, 3).

9 D. I. McCloskey, S. C. Gandevia, E. K. Potter and J. G. Colebatch (1982). Motor commands and judgements about muscular contractions. In *Motor Control in Man* (ed. J. E. Desmedt). New York: Raven Press.

10 Bell (1833), *op cit* (I, 2), 197–98.

11 Gowers and Taylor (1899), *op. cit.* (III, 8), 16–19.

12 W. R. Henderson and G. E. Smyth (1948). Phantom Limbs (cited by Goodwin *et al.* (1972), *loc. cit.* (VI, 3).

13 H. Head (1920). *Studies in Neurology*, Vol. II, 605–08. London: Henry Frowde and Hodder and Stoughton.

14 G. Riddoch (1941). Phantom limbs and body shape. *Brain*, **64,** 197–222.

15 D. I. McCloskey (1978), *loc. cit.* (VI, 4), 803–804.

16 E. Mach (1886), cited by D. I. McCloskey, *loc. cit.* (VI, 4).

17 S. Gandevia (1982). The Perception of Motor Commands or Effort during Muscular Paralysis. *Brain*, **105,** 151–159.

18 A. Brodal (1973), *loc. cit.* (II, 5).

19 D. I. McCloskey (1978), *loc. cit.* (VI, 4), 803.

20 J. C. Rothwell *et al.* (1982), *loc. cit.* (III, 9).

CHAPTER VIII

1 G. Gordon (1978). Editor's Introduction to *Active Touch: the mechanism of recognition of objects by manipulation: a multi-disciplinary approach*. Oxford: Pergamon Press. This book is an invaluable source of information and ideas.

2 C. Bell (1833), *op. cit.* (I, 2), 193, 202–03.

3 C. S. Sherrington (1900). Cutaneous Sensations. In *Textbook of Physiology* (ed. E. A. Schäfer), Vol. 2, 920–1001. Edinburgh and London: Young J. Pentland.

4 R. S. Johansson and Å. B. Vallbo (1983). Tactile sensory coding in the glabrous skin of the human hand. *Trends in Neurosciences*, **6,** 27–32. Cambridge: Elsevier.

5 H. E. Torebjörk, K.-E. Hagbarth and G. Eklund (1978). Tonic finger flexion reflex induced by vibratory activation of digital mechanoreceptors. In *Active Touch* (VIII, 1), 197–203.

6 J. C. Craig (1978). Vibrotactile pattern recognition and masking. In *Active Touch* (VIII, 1), 229–30; J. C. Bliss. Reading machines for the blind. *Ibid.*, 243–48.

7 J. D. Armstrong (1978). The development of tactual maps for the visually handicapped. *Ibid.*, 254.

8 R. S. Johansson and G. Westling (1984). Roles of Glabrous Skin Receptors and Sensorimotor Memory in Automatic Control of Precision Grip When Lifting Rougher or More Slippery Objects. *Exp. Brain Res.*, **56,** 550–64.

9 H. J. Seddon (1975), cited by E. Moberg (1978). *Symposium on the neurologic aspects of plastic surgery* (ed. S. Fredricks and G. S. Brody). Chap. 4: Sensibility in reconstructive limb surgery. St Louis: Mosby.

10 E. Moberg (1964). Aspects of sensation in reconstructive surgery of the upper extremity. *J. Bone and Joint Surg.*, **46-A,** 817–25.

11 *Id.* (1983), VI, 5.

12 *Id.* (1975). Hand surgery and the development of hand prostheses. *Scand. J. Plast. Reconstr. Surg.*, **9,** 227–30.

13 *Id.* (1973). Biological problems in the evolution of hand prostheses. *Symposium on Practical Surgery of the Hand, Orthopedic Clinics of North America*, **4,** 1161–65.

CHAPTER IX

1 C. D. Marsden (1982). The mysterious motor function of the basal ganglia. *Neurology, New York*, **32,** 514–39; V. B. Brooks and W. T. Thach (1981). Cerebellar control of posture and movement. Chapter 18 in *Handbook of Physiology—The Nervous System II*. Bethesda: American Physiological Society.

2 K. S. Lashley (1950). In search of the engram. *Symp. Soc. exp. Biol.*, No. 4, 454–83. Cambridge: Cambridge University Press.

3 L. Deecke, P. Scheid and H. H. Kornhuber (1969). Distribution of readiness potential, pre-motion positivity and motor potential of the human cerebral cortex preceding voluntary finger movements. *Expl. Brain Res.*, **7,** 158–68.

Again the deafferented patient re-enters (III, 9). The waves in his case were indistinguishable from those of normal people. And as in normal people the waves continued to increase during the movement. So this continuing increase must be due to the motor activity itself and not to signals coming in from the moving parts.

On the timing of voluntary movement in relation to the intention to move, see D. I. McCloskey (1980). Knowledge about muscular contractions. *Trends in Neuroscience*, **3,** 311–14. Cambridge: Elsevier; B. Libet, C. A. Gleason, E. W. Wright and D. K. Pearl (1983). Time of conscious intention to act in relation to onset of cerebral activity (readiness-potential). *Brain*, **106,** 623–42.

4 D. Denny-Brown (1966). *The cerebral control of movement*. Sherrington Lectures VIII. Liverpool: Liverpool University Press.

5 H. Cushing (1909). A note upon the faradic stimulation of the post-central gyrus in conscious patients. *Brain*, **32,** 44–53.

6 G. Holmes (1927). Disorders of sensation produced by cortical lesions. In *Selected Papers of Gordon Holmes*, ed. C. G. Phillips, 278–92. Oxford: Oxford University Press; M. Critchley (1953). *The Parietal Lobes*. London: Arnold.

7 W. Penfield (1958). *The Excitable Cortex in Conscious Man*. Sherrington Lectures V. Liverpool: Liverpool University Press.

8 J. H. Jackson (1874–1876). On the scientific and empirical investigation of epilepsies. Reprinted in *Selected Writings of John Hughlings Jackson* (ed. J. Taylor), Vol. I, 162–273. See 262–63. London: Hodder and Stoughton (1931).

9 F. M. R. Walshe (1965). The problem of the origin of the pyramidal tract. In *Further Critical Studies in Neurology*, 96–114. Edinburgh: Livingstone.

10 D. G. Lawrence and D. A. Hopkins (1976). The development of motor control in the rhesus monkey: evidence concerning the role of corticomotoneuronal connections. *Brain*, **99,** 235–54.

11 R. Haaxma and H. G. J. M. Kuypers (1975). Intrahemispheric cortical connexions and visual guidance of hand and finger movements in the rhesus monkey. *Brain*, **98,** 239–60.

12 J. M. Orgogozo and B. Larsen (1979). Activation of the supplementary motor area during voluntary movement in man suggests that it works as a supramotor area. *Science*, **206,** 847–50; D. H. Ingvar and L. Philipson (1977). Distribution of cerebral blood flow in the dominant hemisphere during motor ideation and motor performance. *Ann. Neurol.*, **2,** 230–37.

13 C. Brinkman (1981). Lesions in supplementary motor area interfere with a monkey's performance of a bimanual coordination task. *Neuroscience Letters*, **27,** 267–70.

14 C. S. Roy and C. S. Sherrington (1890). On the regulation of the blood-supply of the brain. *J. Physiol. Lond.*, **11,** 85–108.

15 P. E. Roland, B. Larsen, N. A. Lassen and E. Skinhøj (1980). Supplementary motor area and other cortical areas in organisation of voluntary movements in man. *J. Neurophysiol.*, **43,** 118–36.

16 J. H. Jackson (1874–1876), *loc. cit.* (IX, 8), footnote 2, 262–63.

17 A. W. Campbell (1905). *Histological studies on the localisation of cerebral function*. Cambridge: Cambridge University Press.

18 C. J. Heath, J. Hore and C. G. Phillips (1976). Inputs from low-threshold muscle and cutaneous afferents of hand and forearm to areas 3a and 3b of baboon's cerebral cortex. *J. Physiol., Lond.*, **257,** 199–227.

Cytoarchitectonics is an essential part of the daily practice of brain research. There is a need for different groups of workers to agree on the precise definition of the areas on which they wish to compare results; and the sulci and gyri are not so constant as to be reliable guides for the mapping of boundaries. Researches on the cortex of men and monkeys are of diminished value if they lack cytoarchitectonic map-references, determined histologically in the individual brains and not transferred uncritically from published maps. The major boundaries can be distinguished with fair confidence, as by the arrows in Fig. 21—though the old term 'hair-sharp' has now fallen into disrepute.

19 A. J. Rockel, R. W. Hiorns and T. P. S. Powell (1980). The basic uniformity in structure of the neocortex. *Brain*, **103,** 221–44.

20 See, for example, E. G. Jones (1975). Lamination and differential distribution of thalamic afferent within the sensory-motor cortex of the squirrel monkey. *J. Comp. Neurol.*, **160,** 167–204.

21 See, for example, E. G. Jones and S. P. Wise (1977). Size, laminar and columnar distribution of efferent cells in the sensory-motor cortex of monkeys. *J. Comp. Neurol.*, **175,** 391–438.

22 Personal communications from R. E. W. Fyffe and S. Ghosh (Experimental Neurology Unit, The Australian National University) and Jianping Wu (Brain Research Institute, Academia Sinica, Shanghai).

23 L. Sokoloff (1979). Mapping of local cerebral functional activity by measurement of local cerebral glucose utilization with $[^{14}C]$ deoxyglucose. *Brain*, **102,** 653–668.

24 J. Szentágothai (1978). The Ferrier Lecture: The neuron network of the cerebral cortex: a functional interpretation. *Proc. Roy. Soc. Lond. B.*, **201,** 219–48.

25 V. B. Mountcastle (1966). The neural replication of sensory events in the somatic afferent system. In *Brain and Conscious Experience* (ed. J. C. Eccles), 85–115. See 89. New York: Springer-Verlag.

26 D. H. Hubel and T. N. Wiesel (1977). The Ferrier Lecture: Functional architecture of macaque monkey visual cortex. *Proc. Roy. Soc. Lond. B.*, **198,** 1–59; T. A. Woolsey and H. Van Der Loos (1970). The structural organisation of layer IV in the somatosensory region (SI) of mouse cerebral cortex. *Brain Res.*, **17,** 205–42.

27 M. L Colonnier (1966). The structural design of the neocortex. In *Brain and Conscious Experience* (IX, 25), 1–23. See 17; K. L. Chow and A. L. Leiman (1970). The structural and functional organisation of the neocortex. *Neurosciences Research Program.*, **8,** 157–220. See 176.

28 J. Szentágothai, *loc. cit.* (IX, 24).

29 P. Landry and M. Deschênes (1981). Intracortical arborisations and receptive fields of identified ventrobasal thalamocortical afferents to the primary somatic sensory cortex in the cat. *J. Comp. Neurol.*, **199,** 345–71.

30 C. G. Phillips (1961). Some properties of pyramidal neurones of the motor cortex. In *Ciba Foundation Symposium on the Nature of Sleep* (ed. G. E. W. Wolstenholme and M. O'Connor), 4–24. London: Churchill. See his Fig. 7. A PTN at a depth of 4.5 mm from the convexity of a cat's cortex (lying in the posterior wall of the cruciate sulcus) was powerfully inhibited by surface-cathodal pulses applied to the convexity. But without measurements of current spread, we cannot assert that the tangentially-running axons of the inhibitory interneurones were as long as 4.5 mm.

E. Jankowska, Y. Padel and R. Tanaka (1975). The mode of activation of pyramidal tract cells by intracortical stimuli. *J. Physiol. Lond.*, **249,** 617–36. See their Fig. 7. In monkeys, repetitive intracortical microstimulation fired the corticospinal neurones transynaptically, by exciting the incoming thalamocortical axons. The successive corticospinal volleys increased in amplitude, proving that a larger population of CS neurones discharged in response to each successive pulse. Since the CS neurones are all centred on lamina V, their recruitment can be explained only by invoking tangential spread of synaptic excitation; but the tangential distance remains to be measured.

31 C. G. Phillips, S. Zeki and H. B. Barlow (1984), *loc. cit.* (I, 12).

32 C. G. Phillips, T. P. S. Powell and M. Wiesendanger (1971). Projection from low-threshold muscle afferents of hand and forearm to area 3a of baboon's cortex. *J. Physiol. Lond.*, **217,** 419–46; C. J. Heath, J. Hore and C. G. Phillips (1976), *loc. cit.* (IX, 18).

33 H. Sakata and Y. Iwamura (1978). Cortical processing of tactile information in the first somatosensory and parietal association areas in the monkey. In *Active Touch* (VIII, 1), 55–72; V. B. Mountcastle (1984). Central nervous mechanisms in mechanoreceptive sensibility. *Handbook of Physiology—The Nervous System III*, Chapter 18. Bethesda; American Physiological Society. This gives a splendid perspective of the field.

34 M. M. Merzenich, J. H. Kaas, J. Wall, R. J. Nelson, M. Sur and D. Felleman (1983). Topographic reorganisation of somatosensory cortical areas 3b and 1 in adult monkeys following restricted deafferentation. *Neuroscience*, **8,** 33–55.

35 H. Sakata and Y. Iwamura, *loc. cit.* (IX, 33).

36 V. B. Mountcastle, J. C. Lynch, A. Georgopoulos, H. Sakata and C. Acuna (1975). Posterior parietal association cortex of the monkey: command functions for operations within extrapersonal space. *J. Neurophysiol.*, **38,** 871–908.

37 M. Weinrich, S. P. Wise and K.-H. Mauritz (1984). A neurophysiological study of the premotor cortex in the rhesus monkey. *Brain*, **107,** 385–414; J. Brinkman and R. Porter (1983). Supplementary Motor Area and Premotor Area of Monkey Cerebral Cortex: Functional Organisation and Activities of Single Neurones during Performance of a Learned Movement. In *Motor Control Mechanisms in Health and Disease* (ed. J. E. Desmedt), 393–420. New York: Raven Press.

38 B. Conrad (1978). The motor cortex as a primary device for fast adjust-

ment of programmed motor patterns to afferent signals. In *Cerebral Motor Control in Man: Long Loop Mechanisms*, (*op. cit.*, IV, 16), 123–40.

39 A. S. F. Leyton and C. S. Sherrington (1917), *loc. cit.* (I, 14).

40 Phillips, C. G. (1971). Evolution of the corticospinal tract in primates with special reference to the hand. *Proc. 3rd Int. Congr. Primat. Zurich*, Vol. 2, 2–23. Basel: Karger.

41 E. V. Evarts (1967). Representation of movements and muscles by pyramidal tract neurones of the precentral motor cortex. In *Neurophysiological Basis of Normal and Abnormal Motor Activities* (eds. M. D. Yahr and D. P. Purpura), 215–53. New York: Raven Press; R. N. Lemon, J. A. Hanby and R. Porter (1976). Relationship between the activity of precentral neurones during active and passive movements in conscious monkeys. *Proc. Roy. Soc. Lond. B.*, **194,** 341–73.

42 R. Porter and M. McD. Lewis (1975). Relationship of neuronal discharges in the precentral gyrus of monkeys to the performance of arm movements. *Brain Res.*, **98,** 127–42; P. D. Cheney and E. E. Fetz (1980). Functional classes of primate corticomotoneuronal cells and their relation to active force. *J. Neurophysiol.*, **44,** 773–791.

43 J. Paillard (1955). *Réflexes et Regulations d'Origine Proprioceptive chez l'Homme*. Paris: Librairie Arnette, drew attention to the disproportion between the narrow localization of a motor act and the very widespread facilitation of motoneurones which can be detected in advance of it; J.-M. Coquery and M. Coulmance (1971). Variations d'amplitude des réflexes monosynaptiques avant un mouvement volontaire. *Physiology and Behaviour*, **6,** 65–9.

44 R. Porter and R. B. Muir (1971). The meaning for motoneurones of the temporal pattern of natural activity in pyramidal tract neurones of conscious monkeys. *Brain Res.*, **34,** 127–42.

45 M. Kato and J. Tanji (1972). Conscious control of motor units of human finger muscles. In *Neurophysiology Studied in Man* (ed. G. G. Somjen). Amsterdam: Excerpta Medica.

46 E. E. Fetz and D. V. Finocchio (1972). Operant conditioning of isolated activity in specific muscles and precentral cells. *Brain Res.*, **40,** 19–24.

47 R. N. Lemon (1981). Functional properties of monkey motor cortex neurones receiving afferent input from the hand and fingers. *J. Physiol. Lond.*, **311,** 497–519. See his Fig. 8.

48 C. G. Phillips (1969). The Ferrier Lecture: Motor apparatus of the baboon's hand. *Proc. Roy. Soc. Lond. B.*, **173,** 141–74.

49 E. V. Evarts (1967), *loc. cit.* (IX, 41), Fig. 7.

50 C. D. Marsden, P. A. Merton, H. B. Morton and J. Adam (1977). The effect of posterior column lesions on servo responses from the human long thumb flexor. *Brain*, **100,** 185–200.

51 *Id.* (1977). The effect of lesions of the sensorimotor cortex and the capsular pathways on servo responses from the human long thumb flexor. *Ibid.*, **100,** 503–26. In 1 case out of 15, servo-type responses were *increased*,

suggesting disinhibition of the loop or loops. In rare cases of myoclonic epilepsy, in which the precentral cortical mediation of the jerks is well established, the latency of a reflexly-evoked jerk is about 50 ms. In one such case, in which the adequate stimulus for the jerks was stretch of muscle, the main lesions found post-mortem were in the cerebellum (C. D. Marsden, P. A. Merton and H. B. Morton (1972). *Servo action in human voluntary movement. Nature, Lond.*, **238,** 140–43). If the cerebellum is the gatekeeper of the loop or loops, it had left the gates wide open. *See also* J. A. Obeso, J. C. Rothwell and C. D. Marsden (1985). The spectrum of cortical myoclonus: from focal reflex jerks to spontaneous motor epilepsy. *Brain*, **108,** 193–224.

52 E. G. Jones and S. P. Wise (1977), *op. cit.* (IX, 21).

53 M. Wiesendanger and T. S. Miles (1982). Ascending pathway of low-threshold muscle afferents to the cerebral cortex and its possible role in motor control. *Physiol. Rev.*, **62,** 1234–70. That muscle afferent input from the hand reaches the human cortex has been proved by S. C. Gandevia, D. Burke and B. McKeon (1984). The projection of muscle afferents from the hand to cerebral cortex in man. *Brain*, **107,** 1–13.

54 E. V. Evarts and J. Tanji (1974). Gating of motor cortex reflexes by prior instruction. *Brain Res.*, **71,** 479–94; E. V. Evarts (1985). Transcortical Reflexes: their properties and functional significance. In *Hand Function and the Neocortex*, ed. A. W. Goodwin and I. Darian-Smith. *Exp. Brain Res.*, Suppl. **10,** 130–54; P. D. Cheney and E. E. Fetz (1984). Corticomotoneuronal cells contribute to long-latency stretch reflexes in the rhesus monkey. *J. Physiol. Lond.*, **349,** 249–72.

55 V. B. Brooks and W. T. Thach (1981), *loc. cit.* (IX, 1), 904–06.

56 J. T. Murphy, H. C. Kwan, W. A. MacKay and Y. C. Wong (1978). Spatial organisation of precentral cortex in awake primates. III Input–output coupling. *J. Neurophysiol.*, **41,** 1132–39.

57 E. V. Evarts and C. Fromm (1978). The pyramidal tract neuron as summing point in a closed-loop control system in the monkey. In *Cerebral Motor Control in Man: Long Loop Mechanisms* (*op. cit.*, IV, 16), 56–69. The attenuated response of PTNs in *ballistic* movements may be due, as suggested already (Note IV, 3) to the collateral action of a large synchronized corticospinal volley, inhibiting the somatic relay nuclei so that the afferent volley evoked by the ballistic movement is blocked before it can get to the PTN.

58 P. B. C. Matthews (1984). Evidence from the use of vibration that the human long-latency stretch reflex depends upon spindle secondary afferents. *J. Physiol. Lond.*, **348,** 383–415; *Id.*, The contrasting stretch reflex responses of the long and short flexor muscles of the human thumb. *J. Physiol. Lond.*, **348,** 545–58. These two papers deserve very careful study. See especially ‘A new argument against the transcortical hypothesis based on teleological considerations’, 556–57.

59 D. J. Tracey, B. Walmsley and C. Brinkman (1980). ‘Long-loop’ reflexes can be obtained in spinal monkeys. *Neurosci. Lett.*, **18,** 59–65.

60 G. Eklund, K.-E. Hagbarth, J. V. Hägglund and E. U. Wallin (1982). The 'late' reflex responses to muscle stretch: the 'resonance hypothesis' *versus* the 'long loop' hypothesis. *J. Physiol. Lond.*, **326,** 79–90.

61 A. Berardelli, B. L. Day, C. D. Marsden and J. C. Rothwell (1985). Habituation of the human long latency stretch reflex. *J. Physiol. Lond.*, Proceedings of the Physiological Society; F. A. Lenz, W. G. Tatton and R. R. Tasker (1983). Electromyographic response to displacement of different forelimb joints in the squirrel monkey. *J. Neuroscience*, **3,** 783–94; *Id.* The effect of cortical lesions on the electromyographic response to joint displacement in the squirrel monkey forelimb. *Ibid.*, **3,** 795–805.

62 A fuller account of this subject can be found in *Corticospinal Neurones*, *op cit.* (II, 1), Chapter 5.

63 The powerful method introduced by H. Asanuma and H. Sakata (1967). Functional organisation of a cortical efferent system examined with focal depth stimulation in cats. *J. Neurophysiol.*, **30,** 35–54, and progressively refined and elaborated by Asanuma and his colleagues in many subsequent researches.

64 I. Rosén and H. Asanuma (1972). Peripheral afferent inputs to the forelimb area of the monkey motor cortex: input–output relations. *Expl. Brain Res.*, **14,** 257–73; R. N. Lemon and R. Porter (1976). Afferent input to movement-related precentral neurones in conscious monkeys. *Proc. Roy. Soc. Lond. B.*, **194,** 313–39; R. N. Lemon (1981). Functional properties of monkey motor cortex neurones receiving afferent input from the hand and fingers. *J. Physiol. Lond.*, **311,** 497–519.

65 D. R. Humphrey and W. S. Corrie (1978). Properties of pyramidal tract neuron system within a functionally defined subregion of primate motor cortex. *J. Neurophysiol.*, **41,** 216–43.

66 C. E. Catsman-Berrevoets and H. G. J. M. Kuypers (1979). Differences in distribution of corticospinal axon collaterals to thalamus and midbrain tegmentum in cat, demonstrated by means of double retrograde fluorescent labelling of cortical neurons. *Neuroscience Letters*, Suppl. 3, S133.

67 E. E. Fetz and P. D. Cheney (1980). Post-spike facilitation of forelimb muscle activity by primate corticomotoneuronal cells. *J. Neurophysiol.*, **44,** 751–72.

68 R. N. Lemon, G. W. H. Mantel and R. B. Muir (1984). Organisation of projections from single motor cortex neurones to muscles of the forearm in the conscious monkey. *J. Physiol. Lond.*, **349,** 27P; R. N. Lemon (1983). *Methods for neuronal recording in conscious animals.* IBRO Handbook Series: *Methods in the Neurosciences*, Vol. 4. Chichester: Wiley. See 106–108 and 138–140 (The problem of causality).

69 Y. Shinoda, J.-I. Yokota and T. Futami (1981). Divergent projection of individual corticospinal axons to motoneurons of multiple muscles in the monkey. *Neuroscience Letters*, **23,** 7–12; D. G. Lawrence, R. Porter and S. J. Redman (1985). Corticomotoneuronal synapses in the monkey: light

microscopic localisation upon motoneurons of intrinsic muscles of the hand. *J. Comp. Neurol.*, **232,** 499–510.

Evidence of collateral branching of CS axons in motoneurone pools of different leg muscles was obtained in antidromic electroanatomical experiments by H. Asanuma, P. Zarzecki, E. Jankowska, T. Hongo and S. Marcus (1979). Projection of individual pyramidal tract neurones to lumbar motor nuclei of the monkey. *Exp. Brain Res.*, **34,** 73–89.

70 C. Brinkman, R. Porter and J. Norman (1983). Plasticity of motor behaviour in monkeys with crossed forelimb nerves. *Science*, **220,** 438–40.

71 E. Moberg (1983); *loc. cit.* (VI, 5), 6–7.

72 F. C. Bartlett (1947). The measurement of human skill. *Brit. Med. J.*, **1,** 835–38 and 877–80. See top of page 877.

73 H. Kottenhof (1957). Space perception with experimental spectacles. *Acta Psychol.*, **13,** 79–97.

NOTE ADDED IN PROOF

A new book, *Hand Function and the Neocortex*, edited by A. W. Goodwin and I. Darian-Smith, contains the papers presented at a Satellite Symposium which Drs Goodwin and Darian-Smith had organized at the University of Melbourne in August 1983 in association with the 29th Congress of the International Union of Physiological Sciences at Sydney. 'The issue we hoped to highlight at this meeting was the remarkable interdependence of the sensory and motor functions of the hand and the neural processes mediating them'. The book can be warmly recommended to readers with an appetite for fresh ideas and new experiments.

Reference

Hand Function and the Neocortex (eds. A. W. Goodwin and I. Darian-Smith). Experimental Brain Research, Supplementum **10,** 312 pp. Berlin and Heidelberg: Springer-Verlag, 1985.

Index of names

n indicates a reference, by chapter and note number, to a name appearing in the Notes.